WASTE AND WASTE MANAGEMENT

ENHANCED ELECTROCHEMICAL ADVANCED OXIDATION PROCESSES FOR WASTEWATER SLUDGE STABILIZATION AND REUSE

WASTE AND WASTE MANAGEMENT

Additional books in this series can be found on Nova's website
under the Series tab.

Additional e-books in this series can be found on Nova's website
under the e-book tab.

WASTE AND WASTE MANAGEMENT

ENHANCED ELECTROCHEMICAL ADVANCED OXIDATION PROCESSES FOR WASTEWATER SLUDGE STABILIZATION AND REUSE

GAGIK BADALIANS GHOLIKANDI
EDITOR

nova publishers
New York

Library of Congress Cataloging-in-Publication Data

ISBN: 978-1-63482-089-9

Published by Nova Science Publishers, Inc. † New York

CONTENTS

PREFACE

Sludge management, especially waste activated sludge stabilization and reuse is a fundamental area of concern across the conventional wastewater treatment plants, i.e., the proper waste sludge treatment and disposal technologies should be investigated and provided as a part of the overall waste sludge management concept.

The growing quantities of waste sludge generated in municipal and industrial wastewater treatment plants containing various persistent organics and other contaminants require novel treatment technologies that are capable of achieving significant removal efficiencies and producing reusable sludge products.

The purpose is to increase the waste activated sludge's reuse potential by improving the related quality.

As a response, a large variety of advanced oxidation processes (AOPs) have been developed to remove the present contaminants from wastewaters and limited cases of waste sludge. Multidisciplinary researches have been carried out to investigate and evaluate various related topics such as process fundamentals, determination of system kinetics and mechanisms, exploration of novel materials, and process modeling, integration and scale up. In this regard, electrochemical advanced oxidation processes (EAOPs) have emerged as novel environmental-friendly and effective treatment technologies for the elimination of several organic contaminants. Considerable validation of these methods has been performed at both the bench-scale and pilot-scale. Although a promising new technology, the mechanisms involved in the oxidation of organic compounds during electrochemical advanced oxidation processes and the corresponding environmental impacts have not been completely addressed until now.

This book aims at the electrochemical advanced oxidation processes occurrence from different points of view, describing its related technologies and techniques, providing an assessment of the development and efficiency, and highlighting various aspects of waste activated sludge stabilization and reuse accordingly in five chapters. In the first chapter a brief review to waste sludge stabilization and reuse has been presented. The second chapter provides an overview to advanced oxidation processes. The third chapter describes the various electrochemical advanced oxidation methods. Chapter four presents and discusses the own experimental investigations results employing the Fered-Fenton EAOP. The modeling of the results of the own experimental studies results by means of numerical method and software tools has been performed in the fifth chapter.

The main goal of this book is to gather different updated viewpoints according to the electrochemical advanced oxidation processes and to provide the own experimental studies results accordingly in order to present students, researchers, managers and engineers with useful knowledge in this regard.

Assoc. Prof. Dr. Gagik Badalians Gholikandi
Editor
Faculty of Water Engineering and Environment,
Shahid Beheshti University, A.C., Tehran, Iran
g.badalians@yahoo.com

In: Enhanced Electrochemical Advanced Oxidation ... ISBN: 978-1-63482-089-9
Editor: Gagik Badalians Gholikandi

Chapter 1

WASTE SLUDGE STABILIZATION AND REUSE

Gagik Badalians Gholikandi** and Hamed Rasouli Sadabad
Faculty of Water Engineering and Environment,
Shahid Beheshti University, A.C., Tehran, Iran

INTRODUCTION

Attention to the environmental issues has gradually grown in recent decades considering economic, ecological, and social responsibilities (Brassel-Day et al., 2014). In this regard, it is important to create a circulation of different matters and to recycle the valuable ingredients of waste materials, i.e., agricultural and industrial wastes, wastewater, and waste sludge related.

The waste sludge possesses the potential to notably contaminate the environment if it is not disposed of accurately. The quantity of the waste sludge resulted from wastewater treatment, especially under aerobic conditions (i.e., activated sludge method) is considerable containing a high concentration of different compounds accumulated. Moreover, the conventional waste sludge treatment and disposal methods account for 25-65% of the total operation costs (Liu 2003). Therefore, it has become an important issue related to wastewater treatment plants of economic-technical operation purposes (Appels et al., 2008; Weemaes et al., 1998). Sustainable sludge handling is defined as a method that meets requirements of efficient recycling of resources without supply of harmful materials to humans or the environment (CEC-CD 1986). The significant growth of waste sludge produced quantity and the proper handling are the major challenges accordingly (Valls and Vazquez, 2001; Fytili and Zabaniotou, 2008; Jin et al., 2014). Consequently, the handling methods and disposal techniques, i.e., waste sludge treatment and reuse are a matter of great concern and further efforts at investigation have been made accordingly. Several promising systems were investigated and carried out (Elbeshbishy et al., 2014; Guedes et al., 2014; Ikumi et al., 2014; Magdziarz and Werle 2014; Mayer and Edwards 2014; Martinez-Garcia et al., 2014; Rada et al., 2014; Tang et al., 2014; Wang et al., 2014; Zubrowska-Sudol and Walczak 2014; Caniani et al., 2013; Chen et al., 2013; Donatello and Cheeseman 2013; Hamid and Eskicioglu 2013;

* Corresponding author: Hakimieh- Tehranpars, Shahid Abbaspour Blvd., PO Code 16589-53571, PO Box 16765-1719, Tehran, Iran. Email: g.badalians@yahoo.com

Lu et al., 2013; Valderrama et al., 2013; Yang et al., 2013; Chen and Yan 2012; Dhar et al., 2012; Gustavsson and Engwall 2012; Kelessidis and Stasinakis 2012; Rodriguez et al., 2012; Bernardo and Maschio 2011; Hait and Tare 2011; Khursheed and Kazmi, 2011; Lu et al., 2011; Stefanakis et al., 2011; Wan et al., 2011; Barar et al., 2010; Fountoulakis et al., 2010a, b; Lederer and Rechberger 2010; Rughoonundun et al., 2010; Wett et al., 2010; Xu et al., 2010; Zhao et al., 2010; Alonso et al., 2009; Toreci et al., 2009; Gross et al., 2008; Garces et al., 2008; Liu et al., 2008; Singh and Agrawal 2008; Banegas et al., 2007; Bertin et al., 2007; Hsie et al., 2007; Polat et al., 2007; Aarab et al., 2006; Batchelor 2006; Maranon et al., 2006; Krupp et al., 2005; Mahmoud et al., 2004; Sanchez-Monedero et al., 2004; Monzo et al., 2003; Tiehm et al., 2001).

Sludge is generated and collected at different stages in the wastewater treatment plant and can therefore be known as primary sludge from the mechanical treatment, secondary sludge or biological surplus sludge from the biological treatment or in some cases tertiary sludge or chemical sludge from the post-precipitation stage. The sludge resulted from the wastewater treatment process is usually liquid or semisolid liquid and contains 0.25-12% solids by total weight which is dependent on the treatment method and type of treated wastewater (Metcalf and Eddy 2003, Badalians Gholikandi et al., 2014).

The wastewater sludge contains various classes of inorganic and organic substances in both liquid and solid phases. The liquid phase of sludge contains dissolved substances, which may be organic in the form of carbohydrates and fatty acids or inorganic salts such as ammonium. The solid phase contains organic and inorganic substances, too. Living organisms, their decomposition products and other organic compounds that have been broken down to variable degrees constitute the organic matter while metals and nutrients are regarded as inorganic matter. Remainders from pharmaceuticals such as antibiotics and hormones in the influent wastewater are other substances that also finally end up in the sludge. The concentration of heavy metals in wastewater sludge may vary extensively, depending on the sludge origins. Thus, metals, living organisms, organic substances and nutrients are all mixed together in sewage sludge (Cenni et al., 2001; Jensen and Jepsen, 2005; Goi et al., 2006; Fytili and Zabaniotou, 2008; Mosquera-Losada et al., 2010; Arulrajah et al., 2011; De la Torre et al., 2011; Gusiatin and Kulikowska 2014; Mailler et al., 2014; Soezen et al., 2014). The ingredients of wastewater sludge can be described briefly as follows:

Water Content (Moisture)

The water content can be divided into two groups, i.e., external liquid which includes capillary liquid and cellular water, and internal liquid which includes absorption, adhesion, contact and pore liquids (Lundin et al., 2004).

Metals

Considering the sources of wastewater, compounds of various metals, i.e., cadmium, copper, iron, calcium, chromium, mercury, nickel, zinc, manganese, lead, cobalt, tin, molybdenum, selenium, arsenic, etc. could be found in the waste sludge. Some of these compounds are harmful heavy metals. Several regulations and standards have been proposed in order to reduce the concentration of these metals and inhibit their discharge into the environment. Heavy metals are principal elements restricting the use of sludge for agricultural purposes. The possible accumulation of the heavy metals in soil and their increasing content in plants are the negative effects accordingly. In recent decades, various methods for the heavy metals determination, removal and recovery have been studied in order to minimize the health risks and to improve the extraction efficiency, and to reach finally the different sludge reuse purposes (Alibhai et al., 1985; Fletcher and Beckett 1987a,b; Lake et al., 1989; Blais et al., 1993; Tyagi et al., 1997; Hsiau and Lo 1998; Krogmann et al., 1999; Madoni et al., 1999; Carbonell-Barrachina et al., 2000; Ito et al., 2000; Renoux et al., 2001; Esteves da Silva and Oliveira 2002; Fuentes et al., 2004; Mantovi et al., 2005; Wang et al., 2005; Babel and Del Mundo Dacera 2006; Goi et al., 2006; Beauchesne et al., 2007; Haroun et al., 2007; Hsie et al., 2007; Matsumoto et al., 2007; Zheng et al., 2007; Bouzid et al., 2008; Han et al., 2008; Wang et al., 2008; Xu et al., 2008; Alonso et al., 2009;Hanay et al., 2009; Lasheen and Ammar 2009; Laurent et al., 2009; Mosquera-Losada et al., 2010; Xu et al., 2010; Shukla et al., 2011; Wang et al., 2011; Chen and Yan 2012; Olofsson et al., 2012; Gerardo et al., 2013; Tella et al., 2013; Wang et al., 2013; Xu et al., 2013; Yang et al., 2013; Gusiatin and Kulikowska 2014; Tang et al., 2014; Weng et al., 2014).

Organic Substances

More than 50% of the solid content of wastewater sludge could be organic compounds but the precise amount related depends on the wastewater treatment technology. Organic substances are a mixture of organic contaminants which may be broken down to varying degrees, microorganisms and their metabolic products. Most of the organic matter is soluble compounds such as amino-acids, lipids, hydrocarbons or proteins. Most of the easily degradable organic substances can be decomposed in the biological stage of wastewater treatment. The compounds that are more resistant to biodegradation will accumulate in the sludge. Aside from the substances already existing in the influent wastewater, some organic compounds added during the wastewater treatment, i.e., the polymers added for dewatering purposes of the sludge, can be found in the waste sludge respectively. The effects of the organic loading rate (OLR) on the performance and stability of anaerobic digestion of municipal biomass waste and waste activated sludge must precisely be investigated in order to gain the process goals. The biochemical methane potential (BMP) is a key parameter accordingly (Matsuo et al., 2001; Aparicio et al., 2009; Lozano et al., 2009; Tuan and Sillanpaeae 2010; Wang et al., 2010; Wan et al., 2011; Liu et al., 2012; Dai et al., 2013; Li et al., 2013; Lorenz et al., 2013; Ikumi et al., 2014; Komilis et al., 2014; Mou et al., 2014; Sadef et al., 2014; Soezen et al., 2014; Torrijos et al., 2014).

Nutrients

Waste sludge contains various nutrients. The four most important nutrients in the sludge are phosphorus, nitrogen, potassium and sulfur, which are all essential elements for growing of plants such as crops (limiting factors in crop yield). The settlement characteristics of activated sludge are determined in relation to the variations in the nutrients, i.e., nitrogen and phosphorus, present in the wastewater. The certain combinations of nutrients cause poor settlement. Nitrogen is a part of proteins and nucleic-acids and is therefore vital to all life forms. It is also the nutrient that the majority of crops need the most. Potassium is essential to all organisms and exists as ion mainly in the cell. Phosphorus is a part of the cell's DNA and plays an important role in the metabolism.

It is essential to photosynthesis and respiration processes. Some amino acids contain sulfur as well as iron-sulfur proteins and thiamine. Phosphorus found in the sludge originates mainly from urine, feces and washing detergents which contain phosphates. Various applications have been investigated to create suitable conditions for considerable sludge reduction and nutrient removal or recovery (Forster 1985; Pitman 1999; Ning et al., 2000; Matsuo et al., 2001; Cofie et al., 2006; Correa et al., 2006; Gerardi 2006; Adam et al., 2009; Tong and Chen 2009; de Guardia et al., 2010; Tian and Lu 2010; Jung and Lovitt 2011; Garcia-Albacete et al., 2012; Gustavsson and Engwall 2012; Wei et al., 2012; Gerardo et al., 2013; Weigand et al., 2013; Guedes et al., 2014; Li et al., 2014).

Biological Community

The sludge contains several kinds of microorganisms; some of them are pathogens. The sludge treatment is also crucial because of the pathogens presence. The microorganisms can be divided into different groups, i.e., bacteria, protozoa, worms and viruses.

The type and quantity of the present species can vary considerably with the wastewater treatment methods employed (Matsuo et al., 2001; Gerardi and Zimmerman 2005; Gerardi 2006; Singh and Agrawal 2008; Wackett 2011). The quality assessment of waste sludge and waste sludge-based products employing biotests has been investigated to evaluate the acute toxicity and genotoxicity accordingly (Kapanen et al., 2013).

The main purpose of various sludge treatment processes is to reduce its volume, to stabilize the sludge, to remove water and to kill pathogenic organisms (Matsuo et al., 2001). In the recent decade, many researchers have investigated recognizing methods of the sludge reclamation such as anaerobic digestion for the methane production, anaerobic fermentation for the hydrogen production, aerobic compost for the fertilizer production, and the catalytic reaction of microorganisms (Logan and Regan, 2006, Logan et al., 2006, Razendal et al., 2008).

Table 1. Typical chemical composition and properties of untreated waste sludge (Metcalf and Eddy, 2003, Fytili and Zabaniotou, 2008)

Item	Untreated sludge	
	Range	Typical
Total dry solids (TS), %	2.0–8.0	5.0
Volatile solids (% of TS)	60–80	65
Grease and fats (% of TS)		
Ether soluble	6-30	-
Ether extract	7-35	-
Protein (% of TS)	20–30	25
Nitrogen (% of TS)	1.5–4	2.5
Phosphorous (P_2O_5, % of TS)	0.8–2.8	1.6
Potash (K_2O, % of TS)	0-1	0.4
Cellulose (% of TS)	8.0–15.0	10
Iron (not as sulfide)	2.0–4.0	2.5
Silica (SiO_2, % of TS)	5.0–20.0	-
Alkalinity (mg/L as $CaCO_3$)	500–1500	600
Organic acids (mg/L as Hac)	200–2000	500
Energy content	10,000–12,500	11000
pH	5.0–8.0	6.0

The decision-makers should consider all the technical, economic and environmental data in the optimum way in relation to sludge handling, treatment and disposal.

A typical processes chain is summarized as follows (Matsuo et al., 2001; Metcalf and Eddy 2003):

- preliminary treatment, i.e., screening, comminuting,
- primary thickening, i.e., gravity, flotation, drainage, belt, centrifuge,
- liquid sludge stabilization, i.e., anaerobic digestion, aerobic digestion, lime addition,
- secondary thickening, i.e., gravity, flotation, drainage, belt, centrifuges and conditioning, i.e., elutriation, chemical, thermal,
- Dewatering, i.e., plate press, belt press, centrifuge, drying bed,
- final treatment, i.e., composting, drying, lime addition, incineration, wet oxidation, pyrolysis, disinfection,
- storage, i.e liquid sludge, dry sludge, compost, ash,
- transportation, i.e road, pipeline, sea,
- final destination , i.e., agriculture, horticulture, forest, reclaimed land, land building, other uses.

WASTE SLUDGE STABILIZATION METHODS

The waste sludge management issues and investigations related have been increased permanently because of widely performance of the wastewater activated sludge treatment technology. The requirements for an effective waste sludge management, i.e., waste activated sludge stabilization and reuse are both a significant improvement of the sludge quality and sludge volume reduction.The current waste sludge stabilization methods are as follows:

Anaerobic Sludge Digestion

Anaerobic sludge digestion is considered an effective process for waste activated sludge and mixed sludge stabilization. The occurrence of this method makes both a significant improvement of the sludge quality, i.e., degradation of various polymeric and monomeric substrates, and the sludge volume reduction while generating energy-rich biogas, i.e., energy recovery, possible (Tomei and Carozza 2014; Tyagi et al., 2014). It will also give a partial hygienisation. The total amount of the sludge is reduced to about 35% (Levantesi et al., 2014). The process has been either described by four metabolic steps, i.e., hydrolysis, acidogenesis, acetogenesis, and methanogenesis, or by three steps, i.e., hydrolysis, volatile acid fermentation, and methane formation respectively (Badalians Gholikandi 2009; Gagliano et al., 2014; Gianico et al., 2014; Masihi et al., 2014). Carbohydrates, fats and proteins are broken down in different steps and finally converted into biogas, i.e., methane gas and carbon dioxide. A variety of bacteria participate in the first stages and the methane formation process is derived by methanogenic archaea (Kroeber et al., 2009; Gagliano et al., 2014). Significant abiotic factors in the anaerobic digester include temperature, pH, moisture content, substrate/carbon source, hydraulic retention time (HRT), solid retention time (SRT), ammonia-nitrogen, volatile fatty acids, total solids, volatile solids, and alkalinity (Mutsao, 2001; Nges and Liu 2010; Lee et al., 2011; Gianico et al., 2014; Masihi et al., 2014). This process is not optimized due to the hydrolysis step's limitation means low yield, high retention time, and high digester volume. The various pretreatment methods of waste activated sludge results in an improved sludge digestion efficiency (Lee and Shoda 2008; Zhang et al., 2009; Erden and Filibeli 2010; Lee and Rittmann 2011; Duan et al., 2012; Dai et al., 2013; Gagliano et al., 2014; Gianico et al., 2014; Masihi et al., 2014;Tomei and Carozza 2014; Tyagi et al., 2014) .

Aerobic Digestion

The stabilization of waste activated sludge is by long-term aeration and consequently, biological destruction of volatile solids possible. The related design standards depend on both the waste sludge characteristics and the disposal technology (Hammer and Hammer Jr. 2008). Comparing to anaerobic digestion, the energy cost of aerobic digestion process is notably higher (Matsuo et al., 2001).

Lime Stabilization

The lime stabilizes the sludge, raises the temperature of the mix to 80-100°C, improves water evaporation and the partial oxidation of the organics present in the sludge (Rodriguez et al., 2012). The biological activity of present microorganisms is effectively terminated when the pH value rises above 12. This fact is the base of lime stabilization process. Enough lime, about 30% of the dry solid content of wastewater sludge should be added in order to ensure that no fermentation takes place in the long run. During the liming process, pathogenic microorganisms are killed efficiently; the dry solid content increases and the handling of the sludge become easier. The lime binds phosphates and heavy metals very well. When quicklime is added after dewatering, a sharp rise in temperature occurs and the high temperature pasteurizes the sludge. If quicklime added to sludge before dewatering, eliminates the odor and modifies its dewatering characteristics (Valderrama et al., 2013).

Composting

Composting is an aerobic process and the microorganisms are mainly mesophilic bacteria (the growth at 25-45°C) or thermophilic bacteria (the growth at temperatures above 45°C). The composting process includes two phases, i.e., active composting and curing. During the first phase the decomposition rate is rather high; the easily degradable compounds such as carbohydrates become decomposed, the temperature raises, and the pathogens are destroyed. Optimum operation conditions are 50 to 60°C for 2 to 3 weeks in order to obtain hygienisation. Compared to active composting phase, the decomposition rate and the temperature during the second phase are lower. Fats, cellulose and other complex organic compounds are broken down during this phase. This process takes several weeks. The degradation process of organic micro-contaminants in waste sludge during aerobic composting at various temperatures has been investigated by means of kinetic models which provided suitable descriptions of the considered process (Sadef et al., 2014). The moisture content in the compost should be around 50 to 60% to facilitate the microbial activity optimally. Water is evaporated but also released through microbial activity. It is also important to have the right ratio between carbon and nitrogen. The optimum ratio is about 25-30:1 (C:N) and will gradually decrease to 10-15:1 during the process when most of the organic carbon is converted to carbon dioxide (Matsuo et al., 2001). One of the usual methods of composting is to lay the mixture of sludge and other solid waste (Amir et al., 2010). The solar reactor can provide more stable heat for the composting process and maintaining the temperature above 55°C for more than 3 days is possible (Chen et al., 2014). Biotests for environmental quality assessment of composted sludge have been investigated and proposed (Kapanen et al., 2013).

Thermal Processes

Thermal processes involve removing the organic part of the sludge and just leaving only the ash component for final disposal. Pyrolysis, gasification, wet (air) oxidation, and combustion processes can be categorized in thermal utilization of the wastewater sludge (Hall 1995). The main goal of thermal processing of the wastewate sludge is the utilization of its potential energy and the minimization of environmental impacts simultaneously (Khiari et al., 2004 ; Dennis et al., 2005). Over the recent decades, various technologies have been developed for the thermal processing of sewage sludge as follows:

1. Combustion: Mono- and co-combustion of the wastewater sludge are common established technologies regarding the thermal processes. Mono-combustion is more predominant. Drying of the sludge, release and combustion of the volatile compounds and the combustion of the high ash content remaining as char are the dominant affecting parameters (Hein et al., 1998). The sludge decomposition during the combustion process takes place in the temperature range between 180°C to 580°C resulting about 70% mass loss (Magdziarz and Werle 2014). Based on the composition of the wastewater sludge, its combustion may be seen as potential source of various contaminants and care should be taken during its disposal (Johanson 1994; Zhao et al., 1994; Ogada and Werther 1996).

2. Wet oxidation: wet oxidation is a thermal liquid-phase process. The effective treatment of aqueous wastes containing high organic pollutants is by this method possible. The oxidation process occurs by molecular oxygen at moderate temperatures equals to 180 to 325°C and pressures equal up to 175 bars. Considering the potential for this oxidation process according to the activated sludge treatment under stoichiometric oxygen conditions, temperature proved to be the significant affecting parameter, followed by retention time and oxygen overpressure. A notable sludge volume reduction and COD removal can be resulted and consequently, effluent liquor with a considerable oxygen demand and stable residual solids is obtainable. During the process, the organic content of sewage sludge is thermally hydrolyzed, oxidized and consequently converted to carbon dioxide, water and nitrogen. The efficiency of the liquid-phase oxidation will be considerably improved by means of various heterogeneous or homogeneous catalysts, i.e., lower energy requirements and much higher oxidation efficiency than the conventional wet air oxidation (Luck 1996; Luck 1999; Khan etal., 1999; Debellefontaine and Foussard 2000; Genc etal., 2002; Levec and Pintar 2007).

Two distinctive regimes, i.e., subcritical and supercritical conditions are determined for the process occurrence (Boon and Thomas 1996). Phosphorous can easily be recovered; also heavy metals can be removed from the ashes of supercritical wet oxidation process (Stendahl and Jafverstorm, 2004). Disregard to its advantages, further efforts at this technology have broken down to reduce the considerable capital cost and to limit various operational difficulties (Kolaczkowski etal., 1999).

3. Pyrolysis: pyrolysis is a thermal treatment process in which the sludge (or biomass) is heated to a temperature of 350-500 °C (in some cases 900 °C) under pressure in the absence of oxygen (Stolarek and Ledakowicz, 2001). The wastewater sludge pyrolysis process is a suitable method for energy savings, materials recovery, and finally gaining a no waste residue solution (Samolada and Zabaniotou 2014).The volatile compounds could be completely released to gaseous product in a free-fall pyrolysis reactor at 1300°C. This temperature is in favor of H_2 and CO in the generated gases. The resulted solid residue contains a large ash amount. The most heating value is in the gaseous product (Zhang et al., 2014). The pyrolysis process consists of three stages, i.e., the evaporation of volatile materials, the primary decomposition of non-volatile components, and the advanced decomposition (Khiari et al., 2004).This process converts the sludge into char, ash, oil, water vapor, and combustible gases. For example, the pyrolysis of paper sludge, i.e., a waste product from the paper and pulp manufacturing industry can provide the possibility to convert this waste to high caloric value fuel, biogas, biooil, and charcoal (Strezov and Evans 2009). This technique releases less pollutant than conventional methods such as incineration or combustion, because in this process hazardous materials, such as heavy metals are concentrated in a solid carbonaceous residue (Menendez et al., 2002). The reactions that take place during pyrolysis process are thermal cracking and condensation reactions. The generated solid and/or gaseous products of the pyrolysis can partially be incinerated and used as heating energy in the pyrolysis process.

Several researches have been focused on the mechanisms of the formation of different fractions during pyrolysis process occurence (Font et al., 1996; Urban and Antal 1982; Conesa et al., 1997; Shen and Zhang, 2002; Menendez et al., 2002; Shinogi and Kanri 2003; Sato et al., 2003). One of the most important areas that can be investigated by these techniques is the effect of polymer flocculation on pyrolysis process. Polymer flocculation is advantageous way to improve sludge dewaterability. The flocculation to charge neutralization point would enhance the rate of thermal pyrolysis at higher heating rates (Chu et al., 2001; Chu et al., 2003).The kinetics model is applied to wastewater sludge pyrolysis process for simulating purposes (Ji et al., 2010).

4. Hydrothermal Treatment (thermal hydrolysis): During this process the sludge is heated as an aqueous phase to high temperatures, normally varying between 120 and 400 °C. The hydrothermal treatment process focuses on disintegrating the sludge and results in a formation and/or accumulation of dissolved products, which makes it possible to recover and recycle valuable compounds of the sludge, such as volatile fatty acids, phosphorous compounds, organic compounds and coagulants (Hansen et al., 2000). Dependent upon the chief purpose of the process, there are several modifications such as the temperature changes (Menendez et al., 2002) which can result significant quantities of useful soluble chemical oxygen demand (COD) and volatile fatty acids (Shanableh and Jomaa 2001). Reductive hydrothermal treatment refers to simple aqueous systems heated and pressurized above 300°C and 100 bar under anoxic and/or reducing conditions (Catallo and Comeaux 2008). Hydrothermal treatment by means of subcritical and supercritical water oxidation processes combines sludge mass and volume reduction with the useful organics generation and consequently, supports the biological nutrient removal process. The subcritical water oxidation process, i.e., temperature equals less than 347°C, gain effective hydrolysis equals more than 95% as COD of the sludge organic solids. However, an incomplete oxidation of the organic compounds of the sluge equals less than 95% as COD occurs. The subcritical water oxidation process also produced COD-rich liquors. Complete organics oxidation of the sludge equals more than 99% as COD, including oxidation of hydrolysis products can be achieved employing supercritical water oxidation process, i.e., the temperature equals more than 374°C (Shanableh 2000). The use of an oxidant, such as hydrogen peroxide is another possible modification method which has a positive effect on the production of volatile fatty acids, depending upon the treatment temperature, treatment time, type of sludge, and amount of oxidant (Rulkens 2008). Volatile fatty acids and other dissolved biodegradable organic compounds can be beneficially utilized as an energy or organic carbon source in the denitrification step of the wastewater treatment process and/or in the anaerobic digestion process. The desired heat energy in order to increase the temperature of the sludge can be recovered and reused by heat exchangers. However, it is not clear what happens to the toxics, especially the toxic organics. To solve the sludge problem completely, an intensive post-treatment process will be necessary.

A kinetic model of the activated sludge thermohydrolysis process has been proposed (Imbierowicz and Chacuk 2012).

Calcium chloride assisted hydrothermal treatment improves the activated sludge dewaterability. An increase in temperature and the dosage of the calcium chloride till 60mg/g dry solid allows a continuous improvement of the sludge dewaterability (Yu et al., 2014).

Hydrothermal treatment combined with mechanical expression is a new method investigated to realize liquid/solids separation improvement with low energy consumption (Wang et al., 2014).

Sludge Reuse and Disposal Methods

In order to obtain a sustainable society, it is important to recycle material and energy as far as possible. By recycling for example nutrient materials, the withdrawal of raw material could be decreased significantly. Waste sludge is a resource of useful materials as well as energy. Various studies have been investigated the wastewater sludge reuse from two major viewpoints, i.e., the reuse of the valuable materials of wastewater sludge, and the recovery of its potential energy content in desired forms (Odegaard et al., 2002).

Waste Sludge As a Source of Advantageous Materials

Wastewater sludge contains both organic and inorganic compounds and noticeable parts of them are valuable materials. There are several potentials to reuse these compounds in a beneficial way. For instance lots of attempts have been put into the manufacturing of useful products by thermal solidification of the inorganic sludge compounds, especially in Japan (Taruya et al., 2002; Okuno et al., 2004). The sludge reuse including direct agricultural application and composting is the predominant final disposal method for sludge management in European Union countries, following by incineration and landfilling (landfilling is the common disposal method only in new member states that joined EU after 2004) (Kelessidis and Stasinakis 2012). Lots of methods have been studied in order to reuse the beneficial materials of sewage sludge as follows:

1. Carbon source for treatment plants (sludge hydrolysis): The sludge hydrolysis process is not a treatment method with the purpose to stabilize or dewater the sludge but instead a way to make use of the organic material in the sludge. Sludge contains hydrocarbons which can be utilized as a carbon source to microorganisms. In some cases when wastewater is being processed at the treatment plant, a carbon source is needed in order to modify the efficiency of an especial step like the biological denitrification step. The carbon source must exist in an easily accessible form so it can be used by the denitrifying bacteria, which are able to reduce nitrate into nitrogen gas. This carbon sourse can be obtained by hydrolyzing the sludge which is practical biologically via enzymes, thermally or chemically. Both raw sludge and/or digested sludge yield relatively high concentrations of dissolved organic substances (Matsuo, 2001).

2. Agricultural application: Wastewater sludge contains nutrients, i.e., nitrogen and phosphorous, resulting especially from nitrification-denitrification process stages in wastewater treatment plant. Employing of sludge as a fertilizer enables the recycling all of its nutrients. This fact gives sludge unique fertilizing benefits because various elements contained in the sludge are essential of growing to plants. However, there are several potential negative consequences of the use of wastewater sludge as fertilizer, i.e., the impact of scattering pathogens, organic contaminants and metals. Treated sludge contains rather stable organic compounds and consequently, the structure modification of the humus layer is possible. This modification has positive effects on the cultivating characteristics of the land due to enhanced pore volume, aggregate stability and creating a porous and fine structure accordingly. Another positive effect of utilizing sludge in the agricultural fields is that a high content of the organic materials improves the retention capacity of minerals and water which reduces the probability of water erosion and surface runoff. On the other hand a prolonged fertilizing with sludge decreases the pH value of the soil. There are various processes involving in the acidification of the soil. This concerns sludge precipitated with iron and not potassium. The main concern is washing out of the sulfate. Sludge contains high concentrations of sulfur, which is not used but can be washed out and acidify the land finally. The sludge may contain also other compounds such as hormones and pharmaceutical substances that require being investigated (Laturnus et al., 2007).

Hygienization of the sludge reduces the number of pathogens and also contribute to higher degradation of the organic compounds. Consequently, less organic substances will be applied to the land. The risks of the existing non-degradable organic compounds in the sludge should be explored and monitored. According to the investigations, these substances do not constitute any risks at short-term level, but on a long-term basis, its accumulation might occur and originate several risks. However, the reuse of wastewater sludge for agricultural purposes confronted by social and technical obstacles (Davis 1996). The development of sludge recycling in agriculture, collates, in a grade degree, with the possibilities to enhance the quality of sludge itself and make the public confident about the mentioned issue (Hudson and Lowe 1996).

3. Utilization of wastewater sludge for the cement kiln and the mortar production: One of the advantageous ways to reuse both inorganic and organic compounds of the wastewater sludge is its use in the Portland cement production (Taruya et al., 2002; Okuno et al., 2004; Rodriguez et al., 2012). In this process, which is also applied both ash or dried sludge can be used. Toxic organic contaminants in the sludge are completely oxidized, and due to the high process temperature, heavy metals are immobilized in the cement. Generally, the amounts of ash or dried sludge that are used as raw materials are only a few percentages of the total amount of the raw materials used during the process. Specific requirements set to the sludge quality for use in the production of Portland cement are a suitably low concentration of phosphate and mercury in the sludge. The resulting wastewater sludge ash from incineration can be situated in controlled landfills or used in construction to improve certain properties of building materials (Kaantee et al., 2004). This material has been used to manufacture bricks, to incorporate into concrete mixtures and as a fine aggregate in mortar. Numerous studies have been carried out in this field. According to the results, it was concluded that the sludge can be utilized in construction materials like Portland cement concrete (Monzo et al., 2003; Okuno et al., 2004). On the other hand there are some legislative and environmental issues related to sludge utilization in mortars. Specifications of the European Committee for Standardization (CEN) relating to the use of fly ash as construction material are included in EN 450 and EN 197-1 for concrete and cement, respectively (Cenni et al., 2001).

4. Wastewater sludge as adsorbent: The production of adsorbents such as activated carbon from waste sludge, i.e., the chemical carbonization of dried wastewater sludge for generating activated carbon has been investigated and carried out. Wastewater sludge in wet or hydrated state has also been considered as an adsorbent (Beeckmans and Ng 1971; Bojner 1973; Humphrey et al., 1974; Artola et al., 2000; Bagreev et al., 2001, 2004; Dabrowski et al., 2005; Wang et al., 2008; Smith 2009).

5. Hydrogen production from wastewater sludge: The use of biomass and waste as sources of energy generation, and/or the base material for the hydrogen production constitutes a particularly attractive alternative instead of fossile fuels.The potential of the wastewater sludge for hydrogen production has been investigated by applying downdraft gasification technique (Aznar et al., 1998; Mathieu and Dubuisson 2002). Wet wastewater sludge can be supposed as one of the most frequent feedstock to generate hydrogen gas (Dogru et al., 2002). Hydrogen gas can also be produced by thermal gasification of biomass. Biohydrogen generation by the fermentation process from waste activated sludge has low hydrogen yield because of the limited amount of carbohydrate needed for the use by the hydrogen-forming bacteria. Microbial electrolysis cells (MECs) are originally developed for biological hydrogen production from wastes. It supports also energetically unfavorable biological and chemical reactions. The short-chain fatty acids accumulated in waste activated sludge fermentation can be adoped as an extra carbon source accordingly. The cascade use of organic compounds in the cells increases the biohydrogen production. The microbial electrolysis cells possess high hydrogen yield from waste activated sludge and few hydrogen sinks. In this regard, further investigations are necessary before large scale application (Liu et al., 2012; Lu et al., 2012; Cusick et al., 2014; Zhang and Angelidak 2014).

Wastewater Sludge As a Source of Energy

There are many sludge management options regarding the generation of various forms of energy like heat, electricity, or biofuel. The most important options are as follows:

- Incineration, co-incineration and co-combustion processes in coal-fired power plants or in the production of cement or building materials
- Production of biogas by anaerobic digestion which can be employed in order to produce energy by combustion reaction
- Direct generation of electrical energy by means of specific microorganisms

The biogas production from wastewater sludge attracts more and more attention worldwide (Ting et al., 2004). Supercritical wet oxidation is a promising innovative technology but is still in the development stage (Rulkens 2008).

1. Incineration, co-incineration and co-combustion processes: Incineration and co-incineration of wastewater sludge currently stays as the most attractive disposal method in Europe (Fytili and Zabaniotou 2008).Within these processes, which are applied in the production of Portland cement on a large scale, toxic organic compounds are destructed and the heavy metals are immobilized in the ash or cement as mentioned above. The sludge must be dewatered prior to incineration as much as possible. Dewatered sludge normally contains a dry solid content of 25%.The calorific value of sludge is around 10 to 20 MJ/kg dry solid. The dry solid content of the sludge should be more than 40% in order be incinerated without auxiliary fuel (Malerius and Werther 2003). Modern fluidized bed incinerators have developed into more and more attractive both in terms of capital and operating costs, in comparison to the conventional multiple hearth type. The advantages of the incineration process include a significant reduction of the sludge volume, destruction of the organic compounds, the possibility of energy recovery, and the minimization of odor generation (Minimi 1997; Khiari 2004). However, the incineration technology does not constitute a complete disposal method since approximately 30% of the solids remain as ash. Particulates and fly ash, nitrogen oxides, carbon dioxide, organic compounds, heavy metals bounded to particles, heavy metals in gaseous form, dioxins, acid gases and volatile organic compounds are examples of the possible emissions (Galbreath and Zygarlicke 1996, Hall etal., 1996; Lin etal., 2002; Malerius and Werther 2003).

Co-combustion of wastewater sludge in cement manufacturing is one of its other applications. The final application of a fuel is based upon several criteria, like its cost and accessibility, energy and ash contents, moisture and volatiles contents (Cenni 2001; Kaantee et al., 2004; Valls and Vazquez 2004). In order to attain lower CO_2 emissions the following practices could be followed:

- Lower clinker content: 50% of CO_2 emissions occur during the process of making the intermediate product like clinker. Declines could be achieved by diluting the clinker content and grinding the cement finer.
- Heighten fuel efficiency: the wet process consumes about twice as much fuel as the precalciner process for each tone of cement. However, shifting the process is so expensive, and unlikely to be justified only by fuel savings.
- Alternative fuels: utilization of waste fuels, biofuels or wastewater sludge in cement manufacturing, substituting partly or fully the conventional fuels, is an action that should be undergo a thorough study as for its energy efficiency.

In order to avoid the high costs of a stand-alone incineration plant for sludge and also to increase the energy recovery efficiency, opportunities have been studied to incinerate dried sludge in coal-fired power plants (Luts et al., 2000). The energy recovery from the incineration of the sludge can often significantly be improved by enhancement of dewatering and/or drying processes of sludge and possibly also by the employing of the low-caloric waste heat from the exhaust gases of the power plant. The utilization of the polluted drying gas in the incineration process can lead to a better total performance of process, too. Co-incineration of wastewater sludge in coal-fired power plants is applied in practice.

2. Wet oxidation processes: Energy recovery by wet oxidation process can take place directly by exchanging heat in the reactor or from its exit flow. Compared with sludge incineration process, the treatment of off-gas of supercritical oxidation process is not complicated, therefore the costs of off-gas treatment is negligible (Shanableh and Jomaa 2001). It is also not essential to dewater the sludge before the oxidation process. The existing inorganics in the treated sludge can simply be eliminated from the water phase as ash (Svanstroem et al., 2004).

3. Gasification process: Technologies which can utilize wastes as fuel have great significance. Theoretically, almost all kinds of organic wastes with moisture content of 5 to 30% can be fulfilled; on the other hand, all kinds of bio-fuels cannot lead to the successful gasification (Costello 1999). It is recognized that fuel characterizations such as surface, size, shape as well as moisture content, volatile matter and carbon content affects gasification. Gasification technology can be applied to convert the sewage sludge into a practicable energy and to diminish the environmental problem (Mathieu 2002). Electricity is an attractive form of energy for its relatively high value, simplicity of distribution and ease of adapting to local or national market specifications. Engines and turbines require further development in order to provide compulsory performance guarantees (Aelterman et al., 2006).

Gasification is thermal process during which carbonaceous content of sewage sludge converted to combustible gas and ash in a completely reducing atmosphere like anaerobic digesters. Compared to incineration, gasification can eliminate the occurrence of problems such as the need for supplemental fuel, emissions of sulfur oxides, nitrogen oxides, heavy metals and fly ash and the probable production of chlorinated dibenzodioxins and dibenzofurans because of the fact that it is a net chemically reductive process. Gasification is formed from a series of complex sequential chemical and thermal sub-processes. The whole process is energetically self-sustaining and typically in steady state conditions; no energy input is necessary (Marrero et al., 2003). During the gasification process, wastewater sludge undergoes complex physical and chemical reactions, starting with the drying or losing of water content. The dried wastewater sludge is then pyrolysed and thermally decomposed. In the final step, condensable and non-condensable vapors and char which have been produced during pyrolysis, go through gasification reactions, where they are simultaneously oxidized and reduced to permanent gases at the reduction zone. In the drying zone, wastewater sludge descends into the gasifier and moisture is evaporated due to the heat generated in the underneath zones. The sludge containing less than 15% moisture loses its moisture completely in this zone (Dogru et al., 2002). The actual process is constituted of a number of repeated cycles of mixing char residue of previous gasification, with moist sludge, drying and gasification. This results to the total conversion of the organic carbon in sludge to combustible gas and mineral residues (Marrero et al., 2003; McAuley et al., 2002; Dogru et al., 2001). The usual combustible gases which have been generated during the gasification are carbon monoxide, hydrogen, methane, ethane, and acetylene. The average usable energy content of the produced gas is about 4 mega joules per cubic meter (Dogru et al., 2002). The combustible gas has such a high quality that can be used for power generation to produce heat for sludge drying (Hamilton 2000; Dogru et al., 2002).

One of the most important gasification process problems is according to the heavy metals. During this process heavy metals existing in sewage sludge are accumulated to the final residue. Potentially, heavy metals are able to be consisted in the different phases of the gasification process, i.e., the char residue in the gasifier, the condensate, and the char filter (Reed et al., 2001).

Direct bioelectricity generation from wastewater, waste and waste activated sludge can be done by various methods such as microbial fuel cells (MFCs). Microbial fuel cell (MFC), which converts the chemical energy of organic matter directly into useful electrical energy is a practicable method to produce electricity from materials containing high loads of organic compounds like sewage sludge and wastewater simultaneous with treatment and disposal processes (Dentel et al.,2004; Logan, 2005; Scott and Murano 2007; Appels 2008; Wang 2010). In contrast to conventional fuel cells, MFCs have particular advantages such as high energy-conversion efficiency and mild reaction conditions such as ambient temperature, normal pressure, and neutral pH. Additionally, MFCs can convert numerous types of organic matters into electricity. These organic compounds comprise simple carbohydrates, such as glucose (Chaudhuri and Lovely 2003), acetate and butyrate (Liu et al., 2005), and complex organic compounds such as those in swine wastewaters (Min et al., 2005), domestic wastewater (Logan, 2005), manure and sewage sludge (Scott and Murano 2007). Classic MFCs have two chambers, one anaerobic anode chamber and one aerobic cathode chamber. In the anode chamber, microbes oxidize provided substrates such as organic matter or biomass and release electrons and protons. The electrons are transported to the cathode through an external circuit and the protons are transferred to the cathode chamber (Mohan et al., 2008). In the cathode chamber, the electrons and protons are consumed by an oxidant which is normally oxygen through redox reactions (Watanabe et al., 2008). Utilizing the organic matters in the wastewater or waste activated sludge to produce electricity by MFCs, makes it possible to lower the treatment cost of wastewater and waste sludge (Logan et al., 2006; Du et al., 2007; Rozendal et al., 2008; Rulkens 2008). Several researchers have studied electricity production from wastewater or wastewater sludge using various kinds of MFCs and related conditions (Kim et al., 2002; Vlyssides and Karlis 2004; He and Angenert 2006; Mohan et al., 2008; Jia et al., 2009; Liu et al., 2009; Xiao and Liu 2009; You et al., 2009; Ahn and Logan 2010; Jiang et al., 2010; Song et al., 2010).

Researches according to waste sludge management are running in various related areas such as characterization methods, disposal, and stabilization processes as well as sludge reuse and reclamation. Among them, the researches on both waste sludge stabilization and reuse purposes, especially useful material extraction and energy generation have attracted more attention. The biological energy generation methods from sludge, i.e., bioenergy are still in the initial research phase. In recent years, novel applications regarding waste sludge stabilization and reuse have been investigated and carried out employing advanced oxidation processes (AOPs), especially electrochemical advanced oxidation processes (EAOPs) (Brillas et al., 2009; Badalians Gholikandi et al., 2014; Sires et al., 2014).

REFERENCES

Aarab, T., Smeyers, M., Remy, M., Godden, B. & Delhaye, J.P. (2006) The storage of sewage sludge: Influence of liming on the evolution of copper chemical fractions, Waste Management 26(9), *Elsevier,* 1024-1032.

Adam, C., Peplinski, B., Michaelis, M., Kley, G. & Simon, F.G. (2009) Thermochemical treatment of sewage sludge ashes for phosphorus recovery, Waste Management 29(3), *Elsevier,* 1122-1128.

Aelterman, P., Rabaey, K., Pham, H.T., Boon, N. &Verstraete, W. (2006) Continuous electricity generation at high voltages and currents using stacked microbial fuel cells, *Environmental Science & Technology,* 40, 3388-3394.

Aelterman, P., Verstraete, W. & Rabaey, K. (2006) Microbial fuel cells: methodology and technology, *Environmental Science & Technology,* 40, 5181-5192.

Ahn, Y., Logan, B.E. (2010) Effectiveness of domestic wastewater treatment using microbial fuel cells at ambient and mesophilic temperatures, *Bioresource Technology*, 101, 469-475.

Alibhai, K.R.K., Mehrotra, I. & Forster, C.F. (1985) Heavy metal binding to digested sludge, Water Research 19(12), the International Water Association (IWA), *Elsevier,* 1483-1488.

Alonso, E., Aparicio, I., Santos, J.L., Villar, P. & Santos, A. (2009) Sequential extraction of metals from mixed and digested sludge from aerobic WWTPs sited in the south of Spain, Waste Management 29(1), *Elsevier,* 418-424.

Amir, S., Jouraiphy, A., Meddich, A., El Gharous, M., Winterton, P. & Hafidi, M. (2010) Structural study of humic acids during composting of activated sludge-green waste: Elemental analysis, FTIR and ^{13}C NMR, Journal of Hazardous Materials 177(1-3), *Elsevier,* 524-529.

Aparicio, I., Santos, J.L. & Alonso, E. (2009) Limitation of the concentration of organic pollutants in sewage sludge for agricultural purposes: A case study in South Spain, Waste Management 29(5), *Elsevier,* 1747-1753.

Appels, L., Baeyens, J., Degreve, J. & Dewil, R. (2008) Principles and potential of the anaerobic digestion of waste-activated sludge, *Prog. Energ. Combus. Sci.,* 34,755-781.

Artola, A., Martin, M., Balaguer, M. & Rigola, M. (2000) Isotherm model analysis for the adsorption of Cd (II), Cu (II), Ni (II), and Zn (II) on anaerobically digested sludge, *Journal of Colloid and Interface Science,* 232 (1), 64-70.

Arulrajah, A., Disfani, M.M., Suthagaran, V. & Imteaz, M. (2011) Select chemical and engineering properties of wastewater biosolids, Waste Management 31(12), *Elsevier,* 2522-2526.

Aznar, M.P., Caballero, M.A., Gil, J., Martin, J.A. & Corella, J. (1998) Commercial steam reforming catalysts to improve biomass gasification with steam oxygen mixtures, 2. Catalytic tar removal, Industrial & Engineering Chemistry Research 37, American Chemical Society (ACS), 2668-80.

Babel, S., Del Mundo Dacera, D. (2006) Heavy metal removal from contaminated sludge for land application: A review, Waste Management 26(9), *Elsevier,* 988-1004.

Badalians Gholikandi, G. (2009), Wastewater Treatment Plant Design, first edition, Ayeej Publisher, Tehran, Iran.

Badalians Gholikandi, G., Masihi, H., Azimipour, M., Abrishami, A. & Mirabi, M. (2014) Optimizing stabilization of waste-activated sludge using Fered-Fenton process and artificial neural network modeling (KSOFM, MLP), Journal of Environmental Science and Pollution Research 21(11), *Springer,* 7177-7186.

Bagreev, A., Adib, F. & Bandosz, T.J. (2001) pH of activated carbon surface as an indication of its suitability for H_2S removal from moist air streams, *Carbon,* 39 (12), 1897-1905.

Bagreev, A., Bandosz, T.J. (2004) Efficient hydrogen sulfide adsorbents obtained by pyrolysis of sewage sludge derivedfertilizer modified with spent mineral oil, *Environmental Science & Technology,* 38 (1), 345-351.

Banegas, V., Moreno, J.L., Moreno, J.I., Garcia, C., Leon, G. & Hernandez, T. (2007) Composting anaerobic and aerobic sewage sludges using two proportions of sawdust, Waste Management 27(10), *Elsevier,* 1317-1327.

Barar, S.K., Verma, M., Tyagi, R.D. & Surampalli, R.Y. (2010) Engineered nanoparticles in wastewater and wastewater sludge – Evidence and impacts, Waste Management 30(3), *Elsevier,* 504-520.

Batchelor, B. (2006) Overview of waste stabilization with cement, Waste Management 26(7), *Elsevier,* 689-698.

Beauchesne, I., Cheikh, R.B., Mercier, G., Blais, J.F. & Ouarda, T. (2007) Chemical treatment of sludge : In-depth study on toxic metal removal efficiency, dewatering ability and fertilizing property preservation, Water Research 41(9), the International Water Association (IWA), *Elsevier,* 2028-2038.

Beeckmans, J.M., Ng, P.C. (1971) Pyrolyzed sewage sludge – its production and possible utility, *Environmental Science & Technology,* 5 (1), 69-71.

Bernardo, E., Maschio, R.D. (2011) Glass-ceramics from vitrified sewage sludge pyrolysis residues and recycled glasses, Waste Management 31(11), *Elsevier,* 2245-2252.

Bertin, L., Capodicasa, S., Occulti, F., Girotti, S., Marchetti, L. & Fava, F. (2007) Microbial processes associated to the decontamination and detoxification of a polluted activated sludge during its anaerobic stabilization, Water Research 41(11), the International Water Association (IWA), *Elsevier,* 2407-2416.

Blais, J.F., Tyagi, R.D. & Auclair, J.C. (1993) Bioleaching of metals from sewage sludge : Microorganisms and growth kinetics, Water Research 27(1), the International Water Association (IWA), *Elsevier,* 101-110.

Bojner, G. (1973) Thermal and activated carbon treatment, *Water and Waste Treatment,* 16 (8), 23-25.

Boon, A., Thomas, V. (1996) Resource or rubbish?, *Chemical Engineering Journal, Elsevier,* 25-30.

Bouzid, J., Elouear, Z., Ksibi, M., Feki, M. & Montiel, A. (2008) A study on removal characteristics of copper from aqueous solution by sewage sludge and pomace ashes, Journal of Hazardous Materials 152(2), *Elsevier,* 838-845.

Brassel-Day, A., Reysset, B. & Nyer, R. (2014) Sustainable Development, Sustainable Cities and Military Installations, Springer.

Brillas, E., Sires, I. & Oturan, M.A. (2009) Electro-Fenton Process and Related Electrochemical Technologies Based on Fenton's Reaction Chemistry, Chemical Reviews 109, *American Chemical Society,* (ACS), 6570-6631.

Caniani, D., Masi, S., Mancini, I.M. & Trulli, E. (2013) Innovative reuse of drinking water sludge in geo-environmental applications, Waste Management 33(6), *Elsevier,* 1461-1468.

Carbonell-Barrachina, A.A., Jugsujinda, A., Burlo, F., Delaune, R.D. & Patrick Jr, W.H. (2000) Arsenic chemistry in municipal sewage sludge as affected by redox potential and pH, Water Research 34(1), the International Water Association (IWA), *Elsevier,* 216-224.

Catallo, W.J., Comeaux, J.L. (2008) Reductive hydrothermal treatment of sewage sludge, Waste Management 28(11), *Elsevier,* 2213-2219.

Cenni, R. P., Janish, B., Spliethoff, H. & Hein, K.R.G. (2001) Legislative and environmental issues on the use of ash from coal and municipal sewage sludge co-firing as construction material, Waste Management, 21(1), Elsevier, 17-31.

Chaudhuri, S.K., Lovely, D.R. (2003) Electricity generation by direct oxidation of glucose in mediatorless microbial fuel cells, *Nature Biotechnology,* 21 (10), 1229-1232.

Chen, M., Blanc, D., Gautier, M., Mehu, J. & Gourdon, R. (2013) Environmental and technical assessments of the potential utilization of sewage sludge ashes (SSAs) as secondary raw materials in construction, Waste Management 33(5), *Elsevier,* 1268-1275.

Chen, T., Yan, B. (2012) Fixation and partitioning of heavy metals in slag after incineration of sewage sludge, Waste Management 32(5), *Elsevier,* 957-964.

Chen, Y., Yu, F., Liang, S., Wang, Z., Liu, Z. & Xiong, Y. (2014) Utilization of solar energy in sewage sludge composting: Fertilizer effect and application, Waste Management 34(11), *Elsevier,* 2014-2021.

Chu, C.P., Lee, D. & Chang, C.Y. (2001) Thermal pyrolysis characteristics of polymer flocculated waste activated sludge, Water Research 35(1), the International Water Association (IWA), *Elsevier,* 49-56.

Cofie, O.O., Agbottah, S., Strauss, M., Esseku, H., Montangero, A., Awuah, E. & Kone, D. (2006) Solid-liquid separation of faecal sludge using drying beds in Ghana : Implications for nutrient recycling in urban agriculture, Water Research 40(1), the International Water Association (IWA), *Elsevier,* 75-82.

Commission of European Communities (1986) Council Directive 86/278/EEC of 4 July 1986 on the protection of the environment, and in particular of the soil, when sewage sludge is used in agriculture.

Conesa, J.A., Font, R., Marcilla, A., Caballero, J.A. (1997) Kinetic model for the continues pyrolysis of two types of polyethylene in a fluidized bed reactor, *J. An. Appl. Pyr.,* 40-41,419-431.

Correa, R.S., White, R.E. & Weatherly (2006) Effect of compost treatment of sewage sludge on nitrogen behavior in two soils, Waste Management 26(6), *Elsevier,* 614-619.

Costello, R. (1999) An overview of the US Department of Energy's biomass power program, In: Power production from biomass III, gasification & pyrolysis R&D&D for industry, volume 192, edited by K. Sipila and M. Korhonen, VTT symposium, VTT Espoo.

Cusick, R.D., Ullery, M.L., Dempsey, B.A. & Logan, B.E. (2014) Electrochemical struvite precipitation from digestate with a fluidized bed cathode microbial electrolysis cell, Water Research 54, the International Water Association (IWA), *Elsevier,* 297-306.

Dabrowski, A., Podkoscielny, P., Hubicki, Z. & Barczak, M. (2005) Adsorption of phenolic compounds by activated carbon - a critical review, *Chemosphere,* 58 (8), 1049-1070.

Dai, X., Duan, N. & Dong, B. (2013) High-solids anaerobic co-digestion of sewage sludge and food waste in comparison with monodigestions: Stability and performance, Waste Management 33(2), *Elsevier,* 308-316.

Davis, R.D. (1996) The impact of EU and UK environmental pressures on the future of sludge treatment and disposal, Water and Environment Journal 10, Chartered Institution of Water and Environmental Management (CIWEM), 65-69.

Debellefontaine, H., Foussard, J.N. (2000) Wet air oxidation fir the treatment of industrial wastes. Chemical aspects, reactor design and industrial applications in Europe, Waste Management 20(1), *Elsevier,* 15-25.

De Guardia, A., Mallard, P., Teglia, C., Marin, A., Le Pape, C., Launay, M., Benoist, J.C. & Petiot, C. (2010) Comparison of five organic wastes regarding their behavior during composting: Part 2, nitrogen dynamic, Waste Management 30(3), Elsevier, 415-425.

De la Torre, A., Alonso, E., Concejero, M.A., Sanz, P. & Martinez, M.A. (2011) Sources and behaviour of polybrominated diphenyl ethers (PBDEs), polychlorinated dibenzo-p-dioxins and dibenzofurans (PCDD/Fs) in Spanish sewage sludge, Waste Management 31(6), *Elsevier,* 1277-1284.

Dennis, J.S., Lambert, R.J., Milne, A.J., Scott, S.A. & Hayhurst, A.N. (2005) The kinetics of combustion of chars derived from sewage sludge, *Fuel,* 84,117-126.

Dentel, S.K., Strogen, B. & Chiu, P. (2004) Direct generation of electricity from sludges and other liquid wastes, *Water Science & Technology 50* (9), 161-168.

Dhar, B.R., Nakhla, G. & Ray, M.B. (2012) Techno-economic evaluation of ultrasound and thermal pretreatments for enhanced anaerobic digestion of municipal waste activated sludge, Waste Management 32(3), *Elsevier,* 542-549.

Dogru, M., Midilli, A. & Howarth, C.R. (2002) Gasification of sewage sludge using a throated dowdraft gasifier and uncertainty analysis, *Fuel Process Technology,* 75, 55-82.

Donatello, S., Cheeseman, C.R. (2013) Recycling and recovery routes for incinerated sewage sludge ash (ISSA): A review, Waste Management 33(11), *Elsevier,* 2328-2340.

Du, Z., Li, H., Gu, T. (2007) A state of the art review on microbial fuel cells: a promising technology for wastewater treatment and bioenergy, *Biotechnol. Adv.,* 25, 464-482.

Duan, N., Dong, B., Wu, B. & Dai, X. (2012) High-solid anaerobic digestion of sewage sludge under mesophilic conditions: Feasibility study, Bioresource Technology 104, *Elsevier,* 150-156.

Elbeshbishy, E., Aldin, S., Nakhla, G., Singh, A. & Mullin, B. (2014) Impact of alkaline-hydrolyzed biosolids (Lystek) addition on the anaerobic digestibility of TWAS in lab- and full scale anaerobic digesters, Waste Management 34(11), *Elsevier,* 2090-2097.

Erden, G., Filibeli, A. (2010) Improving anaerobic biodegradability of biological sludges by Fenton pre-treatment: Effects on single stage and two-stage anaerobic digestion, Desalination 251, *Elsevier,* 58-63.

Esteves da Silva, J.C.G., Oliveira, C.J.S. (2002) Metal ion complexation properties of fulvic acids extracted from composted sewage sludge as compared to a soil fulvic acid, Water Research 36(13), the International Water Association (IWA), *Elsevier,* 3404-3409.

European Committee of Standardisation (CEN) (1994) EN 450: Fly ash for concrete, definitions, requirements and quality control.

Fletcher, P., Beckett, P.H.T. (1987a) The chemistry of heavy metals in digested sewage sludge-I. Copper (II) complexation with soluble organic matter, Water Research 21(10), the International Water Association (IWA), *Elsevier,* 1153-1161.

Fletcher, P., Beckett, P.H.T. (1987a) The chemistry of heavy metals in digested sewage sludge-II. Heavy metal complexation with soluble organic matter, Water Research 21(10), the International Water Association (IWA), *Elsevier,* 1163-1172.

Font, R., Marcilla, A., Garcia, A.N., Caballero, J.A. & Conessa, J.A. (1995) Kinetic models for the thermal degradation of heterogeneous materials, *J. An. Appl. Pyr.*, 32, 29-39.

Forster, C.F. (1985) Factors involved in the settlement of activated sludge-I: Nutrients and surface polymers, Water Research 19(10), the International Water Association (IWA), *Elsevier,* 1259-1264.

Fountoulakis, M.S., Makridis, L., Pirounaki, E.K., Chroni, C., Kyriacou, A., Lasaridi, K. & Manios, T. (2010a) Fate and effect of linuron and metribuzin on the co-composting of green waste and sewage sludge, Waste Management 30(1), *Elsevier,* 41-49.

Fountoulakis, M.S., Petousi, I. & Manious, T. (2010b) Co-digestion of sewage sludge with glycerol to boost biogas production, Waste Management 30(10), *Elsevier,* 1849-1853.

Fuentes, A., Llorens, M., Saez, J., Soler, A., Aguilar, M., Ortuno, F. (2004) Phytotoxicity and heavy metals speciation of stabilised sewage sludges, Journal of Hazardous Materials 108(3), *Elsevier,* 161-169.

Gagliano, M.C., Braguglia, C.M., Gallipoli, A., Gianico, A. & Rossetti, S. (2014) Microbial diversity in innovative mesophilic/thermophilic temperature-phased anaerobic digestion of sludge, Environmental Science and Pollution Research, Springer.

Galbreath, K.C., Zygarlicke, J.C. (1996) Mercury speciation in coal combustion and gasification flue gases, Environmental Science & Technology 30(8), *American Chemical Society (ACS),* 2421-2426.

Garces, P., Carrion, M.P., Garcia-Alcocel, E., Paya, J., Monzo, J. & Barrachero, M.V. (2008) Mechanical and physical properties of cement blended with sewage sludge ash, Waste Management 28(12), *Elsevier,* 2495-2502.

Garcia-Albacete, M., Martin, A. & Cartagena, M.C. (2012) Fractionation of phosphorus biowastes : Characterisation and environmental risk, Waste Management 32(6), *Elsevier,* 1061-1068.

Genc, N., Yonsel, S., Dagasan, L. & Onar, A.N. (2002) Wet oxidation: a pre-treatment procedure for sludge, Waste Management 22(6), *Elsevier,* 611-616.

Gerardi, M.H. (2006) Wastewater Bacteria, Wastewater Microbiology Series, Wiley.

Gerardi, M.H., Zimmerman, M.C. (2005) Wastewater Pathogens, Wastewater Microbiology Series, Wiley.

Gerardo, M.L., Zacharof, M.P. & Lovitt, R.W. (2013) Strategies for the recovery of nutrients and metals from anaerobically digested dairy farm sludge using cross-flow microfiltration, Water Research 47(14), the International Water Association (IWA), *Elsevier,* 4833-4842.

Gianico, A., Braguglia, C.M., Gallipoli, A. & Mininni, G. (2014) Innovative two-stage mesophilic/thermophilic anaerobic degradation of sonicated sludge: performances and energy balance, Environmental Science and Pollution Research, Springer.

Goi, D., Tubaro, F. & Dolcetti, G. (2006) Anaysis of metals and EOX in sludge from municipal wastewater treatment plants: A case study, Waste Management 26(2), *Elsevier,* 167-175.

Gross, B., Eder, C., Grziwa, P., Horst, J. & Klimmerle, K. (2008) Energy recovery from sewage sludge by means of fluidized bed gasification, Waste Management 28(10), *Elsevier,* 1819-1826.

Guedes, P., Couto, N., Ottosen, L.M. & Ribeiro, A.B. (2014) Phosphorus recovery from sewage sludge ash through an electrodialytic process, Waste Management 34(5), *Elsevier,* 886-892.

Gusiatin, Z.M., Kulikowska, D. (2014) The usability of the I_R, RAC and MRI indices of heavy metal distribution to assess the environmental quality of sewage sludge composts, Waste Management, 34(7), 1227-1236.

Gustavsson, L., Engwall, M. (2012) Treatment of sludge containing nitro-aromatic compounds in reed-bed mesocosms-Water, BOD, Carbon and nutrient removal, Waste Management 32(1), *Elsevier,* 104-109.

Hait, S., Tare, V. (2011) Optimizing vermistabilization of waste activated sludge using vermicompost as bulking material, Waste Management 31(3), *Elsevier,* 502-511.

Hall, J.E. (1995) Sewage sludge production, treatment and disposal in the European Union, *Water and Environment Journal 9,* Chartered Institution of Water and Environmental Management (CIWEM), 335-342.

Hamid, H., Eskicioglu, C. (2013) Effect of microwave hydrolysis on transformation of steroidal hormones during anaerobic digestion of municipal sludge cake, Water Research 47(14), the International Water Association (IWA), *Elsevier,* 4966-4977.

Hammer, M.J., Hammer, M.J.Jr. (2008) Water and Wastewater Technology, sixth edition, Pearson®-Prentice Hall®, New Jersey, USA.

Han, J., Xu, M., Yao, H., Furuuchi, M., Sakano, T. & Kim, H.J. (2008) Influence of calcium chloride on the thermal behavior of heavy and alkali metals in sewage sludge incineration, Waste Management 28(5), *Elsevier,* 833-839.

Hanay, O., Hasar, H. & Kocer, N.N. (2009) Effect of EDTA as washing solution on removing of heavy metals from sewage sludge by electrokinetic, Journal of Hazardous Materials 119(1-3), *Elsevier,* 703-710.

Hansen, B., Karlsson, I., Cassidy, S. & Pettersson, L. (2000) Operational experiences from a sludge recovery plant, Water Science & Technology 41(8), the International Water Association (IWA), 23-30.

Haroun, M., Idris, A. & Syed Omar, S.R. (2007) A study of heavy metals and their fate in the composting of tannery sludge, Waste Management 27(11), *Elsevier,* 1541-1550.

He, Z., Angenert, L.T. (2006) Application of bacterial biocathodes in microbial fuel cells, *Electroanalysis,* 18, 2009-2015.

Hein, K.R.G., Bemtgen, J.M. (1998) EU clean technology-co-combustion of coal and biomass, Fuel Processing Technology 54, *Elsevier,* 159-169.

Hsiau, P.C., Lo, S.L. (1998) Extractabilities of heavy metals in chemically-fixed sewage sludges, Journal of Hazardous Materials 58(1-3), *Elsevier,* 73-82.

Hsie, C.H., Lo, S.L., Chiueh, P.T., Kuan, W.H. & Chen, C.L. (2007) Microwave enhanced stabilization of heavy metal sludge, *Journal of Hazardous Materials,* 139(1), Elsevier, 160-166.

Hudson, J.A., Lowe, P. (1996) Current technologies for sludge treatment and disposal, *Water and Environment Journal 10,* Chartered Institution of Water and Environmental Management (CIWEM), 436-440.

Humphrey, M.F., Dowler, W.L. & Simmons, G.M. (1974) Carbon Wastewater Treatment Process, Intersociety Conference on Environmental Systems, Report No. 74-ENAs-46. ASME/NASA, Seattle, 2-11.

Ikumi, D.S., Harding, T.H. & Ekama, G.A. (2014) Biodegradability of wastewater and activated sludge organics in anaerobic digestion, Water Research 56, the International Water Association (IWA), *Elsevier,* 267-279.

Imbierowicz, M., Chacuk, A. (2012) Kinetic model of excess activated sludge thermohydrolysis, Water Research 46(17), the International Water Association (IWA), *Elsevier,* 5747-5755.

Ito, A., Umita, T., Aizawa, J., Takachi, T. & Morinaga, K. (2000) Removal of heavy metals from anaerobically digested sewage sludge by a new chemical method using ferric sulfate, Water Research 34(3), the International Water Association (IWA), *Elsevier,* 751-758.

Jensen, J., Jepsen, S.E. (2005) The production, use and quality of sewage sludge in Denmark, Waste Management 25(3), *Elsevier,* 239-247.

Ji, A., Zhang, S., Lu, X. & Liu, Y. (2010) A new method for evaluating the sewage sludge pyrolysis kinetics, Waste Management 30(7), *Elsevier,* 1225-1229.

Jia, B., Liu, Z., Li, X., Yang, Y., Yang, Q., Zeng, G., Liu, Y., Liu, Q. & Zheng S. (2009) Electricity production from surplus sludge using microbial fuel cells, *Envrion. Sci.,* 30, 1227-1231.

Jiang, J., Zhao, Q., Wei, L. & Wang K. (2010) Extracellular biological organic matters in microbial fuel cell using sewage sludge as fuel, Water Research 44(7), the International Water Association (IWA), *Elsevier,* 2163-2170.

Jin, L., Zhang, G. & Tian, H. (2014) Current state of sewage treatment in China, Water Research 66, the International Water Association (IWA), *Elsevier,* 85-98.

Johnson, J.E. (1994) Formation and reduction of nitrogen oxides in fluidised bed combustion, Fuel 73(9), *Elsevier,* 1398-1415.

Jung, I.S., Lovitt, R.W. (2011) Leaching techniques to remove metals and potentially hazardous nutrients from trout farm sludge, Water Research 45(18), the International Water Association (IWA), *Elsevier,* 5977-5986.

Kaantee, U., Zevenhoven, R., Backman, R. & Hupa, M. (2004) Cement manufacturing using alternative fuels and the advantages of process modelling, Fuel Processing Technology 85, *Elsevier,* 293-301.

Kapanen, A., Vikman, M., Rajasaerkkae, J., Vitra, M. & Itaevaara, M. (2013) Biotests for environmental quality assessment of composted sewage sludge, Waste Management 33(6), *Elsevier,* 1451-1460.

Kelessidis, A., Stasinakis, A.S. (2012) Comparative study of the methods used for treatment and final disposal of sewage sludge in European countries, Waste Management 32(6), *Elsevier,* 1186-1195.

Khan, Y., Anderson, G.K. & Elliott, D.J. (1999) Wet oxidation of activated sludge, Water Research 33(7), the International Water Association (IWA), *Elsevier,* 1681-1687.

Khiari, B., Marias, F., Zagrouba, F. & Vaxelaire, J. (2004) Analytical study of the pyrolysis process in a wastewater treatment pilot station, Desalination 167, *Elsevier,* 39-47.

Khursheed, A., Kazmi, A.A. (2011) Retrospective of ecological approaches to excess sludge reduction, Water Research 45(15), the International Water Association (IWA), *Elsevier,* 4287-4310.

Kim, S.H., Kim, W.J. & Chung T.H. (2002) Release characteristics of nitrogen and phosphorus in aerobic and intermittent aerobic sludge digestion, *Korean Journal of Chemical Engineering,* 19, 439-444.

Kolaczkowski, S.T., Plucinski, P., Beltran, F.J., Rivas, F.J. & McLurgh, D.B. (1999) Wet air oxidation: a review of process technologies and aspects in reactor design, *Chemical Engineering Journal* 73(2), *Elsevier,* 143-160.

Komilis, D., Kissas, K. & Symeonidis, A. (2014) Effect of organic matter and moisture on the calorific value of solid wastes: An update of the Tanner diagram, Waste Management 34(2), *Elsevier,* 249-255.

Kroeber, M., Bekel, T., Diaz, N.N., Goesmann, A., Jaenicke, S., Krause, L., Miller, D., Runte, K.J., Viehoever, P., Puehler, A. & Schlueter, A. (2009) Phylogenetic characterization of a biogas plant microbial community integrating clone library 16S-rDNA sequences and metagenome sequence data obtained by 454-pyrosequencing, *Journal of Biotechnology,* 142(1), Elsevier, 38-49.

Krogmann, U., Boyles, L.S., Bamka, W.J., Chaiprapat, S. & Martel, C.J. (1999) Biosolids and sludge management, Water Environment Research 71(5), Water Environment Federation (WEF), 692-714.

Krupp, M., Schubert, J. & Widmann, R. (2005) Feasibility study for co-digestion of sewage sludge with OFMSW on two wastewater treatment plants in Germany, Waste Management 25(4), *Elsevier,* 393-399.

Lake, D.L., Kirk, P.W.W. & Lester, J.N. (1989) Heavy metal solid association in sewage sludges, Water Research 23(3), the International Water Association, *Elsevier,* 285-291.

Lasheen, M.R., Ammar, N.S. (2009) Assessment of metals speciation in sewage sludge and stabilized sludge from different Wastewater Treatment Plants, Greater Cairo, Egypt, *Journal of Hazardous Materials,* 164(2-3), Elsevier, 740-749.

Laturnus, F., von Arnold, K. & Gron, C. (2007) Organic contaminants from sewage sludge applied to agricultural soils - false alarm regarding possible problems for food safety? Environmental Science and Pollution Research 14 (1), *Springer,* 53-60.

Laurent, J., Casellas, M. & Dagot, C. (2009) Heavy metals uptake by sonicated activated sludge: Relation with floc surface properties, Journal of Hazardous Materials 162 (2-3), *Elsevier,* 652-660.

Lederer, J., Rechberger, H. (2010) Comparative goal-oriented assessment of conventional and alternative sewage sludge treatment options, Waste Management 30(6), *Elsevier,* 1043-1056.

Lee, H., Shoda, M. (2008) Stimulation of Anaerobic Digestion of Thickened Sewage Sludge by Iron-rich Sludge Produced bt the Fenton Method, Journal of Bioscience and Bioengineering 106(1), The Society for Biotechnology, Japan, 107-110.

Lee, I.S., Rittmann, B.E. (2011a) Effect of low solids retention time and focused pulsed pre-treatment on anaerobic digestion of waste activated sludge, Bioresource Technology 102(3), *Elsevier,* 2542-2548.

Lee, I.S., Parameswaran, P. & Rittmann, B.E. (2011b) Effects of solids retention time on methanogenesis in anaerobic digestion of thickened mixed sludge, Bioresource Technology 102(22), *Elsevier,* 10266-10272.

Levantesi, C., Beimfohr, C., Blanch, A.R., Carducci, A., Gianico, A., Lucena, F., Tomei, M.C. & Mininni, G. (2014) Hygienization performances of innovative sludge treatment solutions to assure safe land spreading, Environmental Science and Pollution Research, Springer.

Levec, J., Pintar, A. (2007) Catalytic wet-air oxidation processes: A review, Catalysis Today 124(3-4), *Elsevier,* 172-184.

Li, X., Xing, M., Yang, J., Zhao, L. & Dai, X. (2013) Organic matter humification in vermifiltration process for domestic sewage sludge treatment by excitation-emission matrix fluorescence and Fourier transform infrared spectroscopy, *Journal of Hazardous Materials,* 261, *Elsevier,* 491-499.

Li, R., Yin, J., Wang, W., Li, Y. & Zhang, Z. (2014) Transformation of phosphorus during drying and roasting of sewage sludge, Waste Management 34(7), *Elsevier,* 1211-1216.

Lin, S.Y., Suzuki, Y., Hatano, H. & Harada, M. (2002) Developing an innovative method, HyPr-Ring, to produce hydrogen from hydrocarbons, Energy Conservation and Management 81, *Elsevier,* 2079-2085.

Liu, W., Huang, S., Zhou, A., Zhou, G., Ren, N., Wang, A. & Zhuang, G. (2012) Hydrogen generation in microbial electrolysis cell feeding with fermentation liquid of waste activated sludgr, International Journal of Hydrogen Energy, *Elsevier,* 13859-13864.

Liu, Z., Li, X., Jia, B., Zheng, Y., Fang, L., Yang, Q., Wang, D. & Zheng, G. (2009) Production of electricity from surplus sludge using a single chamber floating-cathode microbial fuel cell, Water Science & Technology 60, the International Water Association (IWA), 2399-2404.

Liu, X., Liu, H., Chen, J., Du, G. & Chen, J. (2008) Enhancement of solubilization and acidification of waste activated sludge by pretreatment, Waste Management 28(12), *Elsevier,* 2614-2622.

Liu, H., Logan, B.E. (2004) Electricity generation using an air-cathode single chamber microbial fuel cell in the presence and absence of a proton exchange membrane, Environmental Science & Technology 38, American Chemical Society (ACS), 4040-4046.

Liu, X., Wang, W., Shi, Y., Zheng, L., Gao, X., Qiao, W. & Zou, Y. (2012) Pilot-scale anaerobic co-digestion of municipal biomass waste and waste activated sludge in China: Effect of organic loading rate, Waste Management 32(11), *Elsevier,* 2056-2060.

Logan, B.E., Hamelers, B., Rozendal, S., Schroeder, U., Keller, J., Freguia, S., Aelterman, P., Verstraete, W., Rabaey, K. (2006) Microbial fuel cells: methodology and technology, Environmental Science & Technology 40, American Chemical Society (ACS), 5181-5192.

Logan, B.E., Regan, J.M. (2006) Microbial fuel cells - challenges and applications, Environmental Science & Technology 40, American Chemical Society (ACS), 5172-5180.

Logan, B.E. (2005) Simultaneous wastewater treatment and biological electricity generation, Water Science & Technology 52 (1-2), the International Water Association (IWA), 31-37.

Lorenz, H., Fischer, P., Schumacher, B. & Adler, P. (2013) Current EU-27 technical potential of organic waste streams for biogas and energy production, Waste Management 33(11), *Elsevier,* 2434-2448.

Lozano, C.J.S., Mendoza, M.V., de Arango, M.C. & Monroy, E.F.C. (2009) Microbiological characterization and specific methanogenic activity of anaerobic sludges used in urban solid waste treatment, Waste Management 29(2), *Elsevier,* 704-711.

Lu, X., Shih, K. (2011) Phase transformation and its role in stabilizing simulated lead-laden sludge in aluminum-rich ceramics, Water Research 45(16), the International Water Association (IWA), *Elsevier,* 5123-5129.

Lu, X., Shih, K. & Cheng, H. (2013) Lead glass-ceramics produced from the beneficial use of waterworks sludge, Water Research 47(3), the International Water Association (IWA), *Elsevier,* 1353-1360.

Lu, L., Xing, D., Liu, B. & Ren, N. (2012) Enhanced hydrogen production from waste activated sludge by cascade utilization of organic matter in microbial electrolysis cells, Water Research 46(4), the International Water Association (IWA), *Elsevier,* 1015-1026.

Luck, F. (1996) A review of industrial catalytic wet air oxidation processes, Catalysis Today 27(1-2), *Elsevier,* 195-202.

Luck, F. (1999) Wet air oxidation: past, present and future, Catalysis Today 53(1), *Elsevier,* 81-91.

Luts, D., Devoldere, K., Laethem, B., Bartholomeeusen, W. & Ockier, P. (2000) Co-incineration of dried sewage sludge in coal-fired power plants: A case study, Water Science & Technology 42 (9), the International Water Association (IWA), 259-268.

Madoni, P., Davoli, D. & Guglielmi, L. (1999) Response of sOUR and AUR to heavy metal contamination in activated sludge, Water Research 33(10), the International Water Association (IWA), *Elsevier,* 2459-2464.

Magdziarz, A., Werle, S. (2014) Analysis of the combustion and pyrolysis of dried sewage sludge by TGA and MS, Waste Management 34(1), *Elsevier,* 174-179.

Mahmoud, N., Zeeman, G., Gijzen, H. & Lettinga, G. (2004) Anaerobic stabilization and conversion of biopolymers in primary sludge-effect of temperature and sludge retention time, Water Research 38(4), the International Water Association, *Elsevier,* 983-991.

Mailler, R., Gasperi, J., Chebbo, G. & Rocher, V. (2014) Priority and emerging pollutants in sewage sludge and fate during sludge treatment, Waste Management 34(7), *Elsevier,* 1217-1226.

Malerius, O., Werther, J. (2003) Modelling the adsorption of mercury in the flue gas of sewage sludge incineration, Chemical Engineering Journal 96, *Elsevier,* 197-205.

Mantovi, P., Baldoni, G. & Toderi, G. (2005) Reuse of liquid, dewatered, and composted sewage sludge on agricultural land: effects of long-term application on soil and crop, Water Research 39(2-3), the International Water Association (IWA), *Elsevier,* 289-296.

Maranon, E., Castrillon, L., Fernandez, Y. & Fernandez, E. (2006) Anaerobic treatment of sludge from a nitrification-denitrification landfill leachate plant, Waste Management 26(8), *Elsevier,* 869-874.

Marrero, T., McAuley, B., Sutterlin, W., Morris, S. & Manahan, S. (2003) Fate of heavy metals and radioactive metals in gasification of sewage sludge, Waste Management 24, *Elsevier,* 193-198.

Martinez-Garcia, C.G., Olguin, M.T. & Fall, C. (2014) Aerobic stabilization of biological sludge characterized by an extremely low decay rate: Modeling, identifiability analysis and parameter estimation, Bioresource Technology 166, *Elsevier,* 112-119.

Masihi, H., Badalians Gholikandi, G. & Azimipour, M. (2014) The Anaerobic Sludge Digestion Process, In: Methanogenesis: Biochemistry, Ecological Functions, Natural and Engineered Environments, Chapter 6, Edited by G. Badalians Gholikandi, Nova Science Publishers, 149-171.

Mathieu, P., Dubuisson, R. (2002) Performance analysis of biomass gasifier, Energy Conservation and Management 43, *Elsevier,* 1291-1299.

Matsumoto, N., Uemoto, H. & Saiki, H. (2007) Case study of electrochemical metal removal from actual sediment, sludge, sewage and scallop organs and subsequent pH adjustment of sediment for agricultural use, Water Research 41(12), the International Water Association (IWA), *Elsevier,* 2541-2550.

Matsuo, T., Hanaki, K., Takizawa, S. & Satoh, H. (2001) Advances in Water and Wastewater Treatment Technology: Molecular Technology, Nutrient Removal, Sludge Reduction, and Environmental Health, Elsevier.

Menendez, J.A., Inguanzo, M. & Pis, J.J. (2002) Microwave-induced pyrolysis of sewage sludge, Water Research 36(13), the International Water Association (IWA), *Elsevier,* 3261-3264.

Metcalf & Eddy (2003) Wastewater Engineering: Treatment, Disposal and Reuse, McGraw-Hill, New York.

Meyer, T., Edwards, E.A. (2014) Anaerobic digestion of pulp and paper mill wastewater and sludge, Water Research 65(15), the International Water Association (IWA), *Elsevier,* 321-349.

Mills, N., Pearce, P., Farrow, J., Thorpe, R.B. & Kirkby, N.F. (2014) Environmental & economic life cycle assessment of current & future sewage sludge to energy technologes, Waste Management 34(1), *Elsevier,* 185-195.

Minimi, G., Bartolo, Zuccarello, R.D.I., Lotito, V., Spinosa, L. & Di Pinto, A.C. (1997) A design model of sewage sludge incineration plants with energy recovery, Water Science & Technology 36, the International Water Association (IWA), 211-218.

Mohan, S.V., Saravanan, R., Raghavulu, S.V., Mohanakrishna, G. & Sarma, P.N. (2008) Bio-electricity production from wastewater treatment in dual chambered microbialfuel cell (MFC) using selectively enriched mixed microflora-effect of catholyte, Bioresource Technology 99, *Elsevier,* 596-603.

Monzo, J., Paya, J., Borrachero, M.V. & Girbes, I. (2003) Reuse of sewage sludge ashes (SSA) in cement mortars, Waste Management 23, *Elsevier,* 373-381.

Mosquera-Losada, M.R., Munoz-Ferreiro, N. & Rigueiro-Rodriguez, A. (2010) Agronomic characterization of different types of sewage sludge: Policy implications, Waste Management 30(3), *Elsevier,* 492-503.

Mou, Z., Scheutz, C. & Kjeldsen, P. (2014) Evaluating the biochemical methane potential (BMP) of low-organic waste at Danish landfills, Waste Management 34(11), *Elsevier,* 2251-2259.

Nges, I.A., Liu, J. (2010) Effects of solid retention time on anaerobic digestion of dewatered-sewage sludge in mesophilic and thermophilic conditions, Renewable Energy 35(10), *Elsevier,* 2200-2206.

Ning, Z., Patry, G.G. & Spanjers, H. (2000) Identification and quantification of nitrogen nutrient deficiency in the activated sludge process using respirometry, Water Research 34(3), the International Water Association (IWA), *Elsevier,* 3345-3354.

Odegaard, H., Paulsrud, B. & Karlsson, I. (2002) Wastewater sludge as a resource: sludge disposal strategies and corresponding treatment technologies aimed at sustainable handing of wastewater sludge, Water Science & Technology 46 (10), the International Water Association (IWA), 295-303.

Ogada, T., Werther, J. (1996) Combustion characteristics of wet sludge in a fluidised bed, release and combustion of volatiles, Fuel 75(5), *Elsevier,* 617-626.

Olofsson, U., Bignert, A. & Haglund, P. (2012) Time-trends of metals and organic contaminants in sewage sludge, Water Research 46(15), the International Water Association (IWA), *Elsevier,* 4841-4851.

Pitman, A.R. (1999) Management of biological nutrient removal plant sluges-Change the paradigms? Water Research 33(5), the International Water Association (IWA), *Elsevier,* 1141-1146.

Polat, M., Guler, E., Lederman, E. & Cohen, H. (2007) Neutralization of an extremely acidic sludge and stabilization of heavy metals in flyash aggregates, Waste Management 27(4), *Elsevier,* 482-489.

Rada, E.C., Ragazzi, M., Villotti, S. & Torretta, V. (2014) Sewage sludge drying by energy recovering from OFMSW composting: Preliminary feasibility evaluation, Waste Management 34(5), *Elsevier,* 859-866.

Reed, G.P., Dugwell, D.R. & Kandiyoti, R. (2001) Control of trace elements in gasification: distribution to the output streams of a pilot scale gasifier, Energy & Fuels15, American Chemical Society (ACS),794-800.

Renoux, A.Y., Tyagi, R.D. & Samson, R. (2001) Assessment of toxicity reduction after metal removal in bioleached sewage sludge, Water Research 35(6), the International Water Association (IWA), *Elsevier,* 1415-1424.

Rodriguez, N.H., Granados, R.J., Blanco-Varela, M.T., Cortina, J.L., Martinez-Ramirez, S., Marsal, M., Guillem, M., Puig, J., Fos, C., Larrotcha, E. & Flores, J. (2012) Evaluation of a lime-mediated sewage sludge stabilization process. Product characterization and technological validation for tis use in the cement industry, Waste Management 32(3), *Elsevier,* 550-560.

Rozendahl, R.A., Hamelers, H.V.M., Rabaey, K., Keller, J. & Buisman, C.J.N. (2008) Towards practical implementation of bioelectrochemical wastewater treatment, Trends in Biotechnology 26, *Elsevier,* 450-459.

Rughoonundun, H., Granda, C., Mohee, R. & Holtzapple, M.T. (2010) Effect of thermochemical pretreatment on sewage sludge and its impact on carboxylic acids production, Waste Management 30(8-9), *Elsevier,* 1614-1621.

Rulkens, W. (2008) Sewage sludge as a biomass resource for the production of energy: overview and assessment of the various options, Energy & Fuels 22, American Chemical Society (ACS), 9-15.

Sadef, Y., Poulsen, T.G. & Bester, K. (2014) Modeling organic micro pollutant degradation kinetics during sewage sludge composting, Waste Management 34(11), *Elsevier,* 2007-2013.

Samolada, M.C., Zabaniotou, A.A. (2014) Comparative assessment of municipal sewage sludge incineration, gasification and pyrolysis for a sustainable sludge-to-energy management in Greece, Waste Management 34(2), *Elsevier,* 411-420.

Sanchez-Monedero, M.A., Mondini, C., de Nobili, M., Leita, L. & Roig, A. (2004) Land application of biosolids. Soil response to different stabilization degree of the treated organic matter, Waste Management 24(4), *Elsevier,* 325-332.

Sato, S., Lin, S.Y., Suzuki, Y. & Hatano, H. (2003) Hydrogen production from heavy oil in the presence of calcium hydroxide, Fuel 82, *Elsevier,* 561-567.

Scott, K., Murano, C. (2007) A study of a microbial fuel cell battery using manure sludge waste, Journal of Chemical Technology and Biotechnology 82, Society of Chemical Industry (SCI), 809-817.

Shanableh, A. (2000) Production of useful organic matter from sludge using hydrothermal treatment, Water Research 34(3), the International Water Association (IWA), *Elsevier,* 945-951.

Shanableh, A., Jomaa, S. (2001) Production and transformation of volatile fatty acids from sludge subjected to hydrothermal treatment, Water Science & Technology 44 (10), the International Water Association (IWA), 129-136.

Shen, L., Zhang, D.K. (2002) An experimental study of oil recovery from sewage sludge by low-temperature pyrolysis in a fluidised-bed, Fuel 82, *Elsevier,* 465-472.

Shinogi, Y., Kanri, Y. (2003) Pyrolysis of plant, animal and human waste: physical and chemical characterisation of pyrolytic products, Bioresource Technology 90, *Elsevier,* 241-247.

Shukla, O.P., Juwarkar, A.A., Singh, S.K., Khan, S. & Rai, U.N. (2011) Growth responses and metal accumulation capabilities of woody plants during the phytoremediation of tannery sludge, Waste Management 31(1), *Elsevier,* 115-123.

Singh, R.P., Agrawal, M. (2008) Potential benefits and risks of land application of sewage sludge, Waste Management 28(2), *Elsevier,* 347-358.

Sires, I., Brillas, E., Oturan, M.A., Rodrigo, M.A. & Panizza, M. (2014) Electrochemical advanced oxidation processes: today and tomorrow. A review, Journal of Environmental Science and Pollution Research 21(14), *Springer,* 8336-8367.

Smith, K.M., Fowler, G.D., Pullket, S. & Graham, N.J.D. (2009) Sewage sludge-based adsorbents: A review of their production, properties and use in water treatment applications, Water Research 43, the International Water Association (IWA), *Elsevier,* 2569-2594.

Soezen, S., Cokgor, E.U., Insel, G., Tas, D.O., Dulkadiroglu, H., Karaca, C., Filibeli, A., Meric, S. & Orhon, D. (2014) Scientific basis of dissolved organic carbon limitation for landfilling of municipal treatment sludge-Is it attainable and justifiable?, Waste Management 34(9), *Elsevier,* 1657-1666.

Song, L.J., Zhu, N.W., Yuan, H.P., Hong, Y. & Ding J. (2010) Enhancement of waste activated sludge aerobic digestion by electrochemical pre-treatment, Water Research 44, the International Water Association (IWA), *Elsevier,* 4371-4378.

Stefanakis, A.I., Komilis, D.P. & Tsihrintzis, V.A. (2011) Stability and maturity of thickened wastewater sludge treated in pilot-scale sludge treatment wetlands, Water Research 45(19), the International Water Association (IWA), *Elsevier,* 6441-6452.

Stendahl, K., Jaefverstroem, S. (2004) Recycling of sludge with the Aqua Reci process, Water Science & Technology 49 (10), the International Water Association (IWA), 233-240.

Stolarek, P., Ledakowicz, S. (2001) Thermal processing of sewage sludge by drying, pyrolysis, gasification and combustion, Water Science & Technology 44 (10), the International Water Association (IWA), 333-340.

Strezov, V., Evans, T.J. (2009) Thermal processing of paper sludge and characterization of its pyrolysis products, Waste Management 29(5), *Elsevier,* 1644-1648.

Svanstroem, M., Modell, M. & Tester, J. (2004) Direct energy recovery from primary and secondary sludges by supercritical water oxidation, Water Science & Technology 49 (10), the International Water Association (IWA), 201-208.

Tang, Y., Chan, S.W. & Shih, K. (2014) Copper stabilization in beneficial use of waterworks sludge and copper-laden electroplating sludge for ceramic materials, Waste Management 34(6), *Elsevier,* 1085-1091.

Taruya, T., Okuno, N. & Kanaya, K. (2002) Reuse of sewage sludge as raw material of Portland cement in Japan, Water Science & Technology 46 (10), the International Water Association (IWA), 255-258.

Tella, M., Doelsch, E., Letourmy, P., Chataing, S., Cuoq, F., Bravin, M.N. & Saint Macary, H. (2013) Investigation of potentially toxic heavy metals in different organic wastes used to fertilize market garden crops, Waste Management 33(1), *Elsevier,* 184-192.

Tian, Y., Lu, Y. (2010) Simultaneous nitrification and denitrification process in a new Tubificidae-reactor for minimizing nutrient release during sludge reduction, Water Research 44(20), the International Water Association (IWA), *Elsevier,* 6031-6040.

Tiehm, A., Nickel, K., Zellhorn, M. & Neis, U. (2001) Ultrasonic waste activated sludge disintegration for improving anaerobic stabilization, Water Research 35(8), the International Water Association (IWA), *Elsevier,* 2003-2009.

Ting, C.H., Lin, K.R., Lee, D.J. & Tay, J.H. (2004) Production of hydrogen and methane from wastewater sludge using anaerobic fermentation, Water Science & Technology 50 (9), the International Water Association (IWA), 223-228.

Tomei, M.C., Carozza, N.A. (2014) Sequential anaerobic/aerobic digestion for enhanced sludge stabilization: comparison of the process performance for mixed and waste sludge, Environmental Science and Pollution Research, Springer.

Tong, J., Chen, Y. (2009) Recovery of nitrogen and phosphorus from alkaline fermentation liquid of waste activated sludge and application of the fermentation liquid to promote biological municipal wastewater treatment, Water Research 43(12), the International Water Association (IWA), *Elsevier,* 2969-2976.

Toreci, I., Kennedy, K.J. & Droste, R.L. (2009) Evaluation of continuous mesophilic anaerobis sludge digestion after high temperature microwave pretreatment, Water Research 43(5), the International Water Association (IWA), *Elsevier,* 1273-1284.

Torrijos, M., Sousbie, P., Rouez, M., Lemunier, M., Lessard, Y., Galtier, L., Simao, A. & Steyer, J.P. (2014) Treatment of the biodegradable fraction of used disposable diapers by co-digestion with waste activated sludge, Waste Management 34(3), *Elsevier,* 669-675.

Tuan, P.A., Sillanpaeae, M. (2010) Migration of ions and organic matter during electro-dewatering of anaerobic sludge, Journal of Hazardous Materials 173(1-3), *Elsevier,* 54-61.

Tyagi, R.D., Blais, J.F., Meunier, N. & Benmoussa, H. (1997) Simultaneous sewage sludge digestion and metal leaching-effect of sludge solids concentration, Water Research 31(1), the International Water Association (IWA), *Elsevier,* 105-118.

Tyagi, V.K., Lo, S.L. & Rajpal, A. (2014) Chemically coupled microwave and ultrasonic pre-hydolysis of pulp and paper mill waste-activated sludge: effect on sludge solubilisation and anaerobic digestion, Environmental Science and Pollution Research 21, *Springer,* 6205-6217.

Urban, D.L., Antal Jr., M.J. (1982) Study of the kinetics of sewage sludge pyrolysis using DSC and TGA, Fuel 61(9), *Elsevier,* 799-806.

Valderrama, C., Granados, R. & Cortina, J.L. (2013) Stabilization of dewatered domestic sewage sludge by lime addition as raw material for the cement industry: Understanding process and reactor performance, *Chemical Engineering Journal* 232, *Elsevier,* 458-467.

Valls, S., Vazquez, E. (2000) Stabilisation and solidification of sewage sludges with Portland cement, Cement and Concrete Research 31, *Elsevier,* 1671-1678.

Valls, S., Vazquez, E. (2001) Accelerated carbonation of sewage sludge-cement-sand mortars and its environmental impact, Cement and Concrete Research 31, *Elsevier,* 1271-1276.

Vlyssides, A.G., Karlis, P.K. (2004) Thermal-alkaline solubilization of waste activated sludge as a pre-treatment stage for anaerobic digestion, Bioresource Technology 91, *Elsevier,* 201-206.

Wan, C., Zhou, Q., Fu, G. & Li, Y. (2011) Semi-continous anaerobic co-digestion of thickened waste activated sludge and fat, oil and grease, Waste Management 31(8), *Elsevier,* 1752-1758.

Wang, X., Chen, T., Ge, Y. & Jia, Y. (2008) Studies on land application of sewage sludge and its limiting factors, Journal of Hazardous Materials 160(2-3), *Elsevier,* 554-558.

Wang, C., Hu, X., Chen, M.L. & Wu, Y.H. (2005) Total concentration and fractions of Cd, Cr, Pb, Cu, Ni, and Zn in sewage sludge from municipal and industrial wastewater treatment plants, Journal of Hazardous Materials 119(1-3), *Elsevier,* 245-249.

Wang, Q., Jiang, G., Ye, L. & Yuan, Z. (2014) Enhancing methane production from waste activated sludge using combined free nitrous acid and heat pre-treatment, Water Research 63, the International Water Association (IWA), *Elsevier,* 71-80.

Wang, W., Luo, Y. & Qian, W. (2010) Possible solutions for sludge dewatering in China, Frontiers of Environmental Science & Engineering in China 4, 102-107.

Wang, X., Qiu, Z., Lu, S. & Ying, W. (2010) Characteristics of organic, nitrogen and phosphorus species released from ultrasonic treatment of waste activated sludge, *Journal of Hazardous Materials,* 176 (1-3), Elsevier, 35-40.

Wang, X.J., Xu, X.M., Liang, X., Wang, Y., Liu, M., Wang, X., Xia, S.Q., Zhao, J.F., Yin, D.Q. & Zhang, Y.L. (2011) Adsorption of copper (II) onto sewage sludge-derived materials via microwave irradiation, Journal of Hazardous Materials 192(3), *Elsevier,* 1226-1233.

Wang, L., Zhang, L. & Li, A. (2014) Hydrothermal treatment coupled with mechanical expression at increased temperature for excess sludge dewatering : Influence of operating conditions and the process energetics, Water Research 65, the International Water Association (IWA), *Elsevier,* 85-97.

Wang, L., Zheng, Z., Zhang, Y., Chao, J., Gao, Y., Luo, X. & Zhang, J. (2013) Biostabilization enhancement of heavy metals during the vermiremediation of sewage sludge with passivant, *Journal of Hazardous Materials*, 244-245, *Elsevier,* 1-9.

Wang, X., Zhu, N., Yin, B. (2008) Preparation of sludge-based activated carbon and its application in dye wastewater treatment, Journal of Hazardous Materials 153 (1-2), *Elsevier,* 22-27.

Watanabe, K. (2008) Recent developments in microbial fuel cell technologies for sustainable bioenergy, *Journal of Bioscience and Bioengineering* 106, the Society for Biotechnology, Japan, *Elsevier,* 528-536.

Wei, Y., Ji, M., Li, R. & Qin, F. (2012) Organic and Nitrogen removal from landfill leachate in aerobic granular sludge sequencing batch reactors, Waste Management 32(3), *Elsevier,* 448-455.

Weigand, H., Bertau, M., Huebner, W., Bohndick, F. & Bruckert, A. (2013) RecoPhos: Full-scale fertilizer production from sewage sludge ash, Waste Management 33(3), *Elsevier,* 540-544.

Weng, H.X., Ma, X.W., Fu, F.X., Zhang, J.J., Liu, Z., Tian, L.X. & Liu, C. (2014) Transformation of heavy metal speciation during sludge drying: Mechanistic insights, *Journal of Hazardous Materials*, 265, *Elsevier*, 96-103.

Wett, B., Phothilangka, P. & Eladawy, A. (2010) Systematic comparison of mechanical and thermal sludge disintegration technologies, Waste Management, 30(6), 1057-1062.

Xiao, B.Y., Liu, J.X. (2009) Effects of various pretreatments on biohydrogen production from sewage sludge, *Chinese Science Bulletin*, 54, 2038-2044.

Xu, G., Liu, M. & Li, G. (2013) Stabilization of heavy metals in lightweight aggregate made from sewage sludge and river sediment, *Journal of Hazardous Materials*, 260, *Elsevier*, 74-81.

Xu, G.R., Zou, J.L. & Li, G.B. (2008) Solidification and leaching behaviours of Cr^{6+} in sludge ceramsite, *Journal of Hazardous Materials*, 153(3), *Elsevier*, 1031-1035.

Xu, G.R., Zou, J.L. & Li, G.B. (2010) Stabilization of heavy metals in sludge ceramsite, Water Research 44(9), the International Water Association (IWA), *Elsevier*, 2930-2938.

Yang, J., Zhao, C., Xing, M. & Lin, Y. (2013) Enhancement stabilization of heavy metals (Zn, Pb, Cr and Cu) during vermifiltration of liquid-state sludge, Bioresource Technology 146, *Elsevier*, 649-655.

You, S.J., Ren, N.Q., Zhao, Q.L., Kiely, P.D., Wang, J.Y., Yang, F.L., Fu, L. & Peng, L. (2009) Improving phosphate buffer-free cathode performance of microbial fuel cell based on biological nitrification, Biosensors and Bioelectronics 24, *Elsevier*, 3698-3701.

Yu, J., Guo, M., Xu, X. & Guan, B. (2014) The role of temperature and $CaCl_2$ in activated sludge dewatering under hydrothermal treatment, Water Research 50, the International Water Association (IWA), *Elsevier*, 10-17.

Zhang, Y., Angelidaki, I. (2014) Microbial electrolysis cells turning to be versatile technology: Recent advances and future challenges, Water Research 56, the International Water Association (IWA), *Elsevier*, 11-25.

Zhang, H., Banaszak, J.E., Parameswaran, P., Alder, J., Krajmalnik-Brown, R. & Rittmann, B.E. (2009) Focused-Pulsed sludge pre-treatment increases the bacterial diversity and relative abundance of acetoclastic methanogens in a full-scale anaerobic digester, Water Research 43(18), the International Water Association (IWA), *Elsevier*, 4517-4526.

Zhang, L., Xiao, B., Hu, Z., Liu, S., Cheng, G., He, P. & Sun, L. (2014) Tar-free fuel gas production from high temperature pyrolysis of sewage sludge, Waste Management 34(1), *Elsevier*, 180-184.

Zhao, J., Grace, J.R., Lim, J.C., Clive, M.H., Brereton, X. & Legros, R. (1994) Influence of operating parameters on NOx emissions from circulating fluidised bed combustor, Fuel 73 (10), *Elsevier*, 1650-1657.

Zhao, L., Wang, Y., Yang, J., Xing, M., Li, X., Yi, D. & Deng, D. (2010) Earthworm-microorganism interactions: A strategy to stabilize domestic wastewater sludge, Water Research 44(8), the International Water Association (IWA), *Elsevier*, 2572-2582.

Zheng, G.D., Gao, D., Chen, T.B. & Luo, W. (2007) Stabilization of nickel and chromium in sewage sludge during aerobic composting, Journal of Hazardous Materials 142(1-2), *Elsevier*, 216-221.

Zubrowska-Sudol, M., Walczak, J. (2014) Effects of mechanical disintegration of activated sludge on the activity of nitrifying and denitrifying bacteria and phosphorus accumulating organisms, Water Research 61, the International Water Association (IWA), *Elsevier*, 200-209.

In: Enhanced Electrochemical Advanced Oxidation ... ISBN: 978-1-63482-089-9
Editor: Gagik Badalians Gholikandi

Chapter 2

ADVANCED OXIDATION PROCESSES (AOPS): AN OVERVIEW

Gagik Badalians Gholikandi*[*] *and Mina Nili Ardakani
Faculty of Water Engineering and Environment,
Shahid Beheshti University, A.C., Tehran, Iran

INTRODUCTION

Over the last decades, the environmental contamination and the related awareness has been considered as a world-wide issue due to the significant increase in the problems accordingly (Safa and Soucy 2014). Wastewater and the waste sludge, especially municipal and industrial wastewater, landfill leachate, etc. may contain toxic and refractory organic contaminants and its uncontrolled discharge into the environment can cause serious environmental problems. The conventional technologies including biological, physical, chemical, physicochemical, and thermal treatment processes have been employed to remove the considered aqueous contaminants and to control its discharge into the environment. Although biological-chemical, physicochemical or thermal approaches are widely investigated and applied accordingly, there is still new improved approaches needed to overcome the related process disadvantages and to present novel effective techniques (Debellefontaine et al., 1996; Adewuyi 2005a; Kim and Ihm 2011).The familiar conventional methods limitations have encouraged the researchers to develop more efficient and environmental-friendly systems for wastewater treatment and waste sludge stabilization. Since 1990, a lot of researches have been addressed to this aim pointing out the prominent role of a special class of oxidation techniques defined as advanced oxidation processes (AOP) which usually operate at or near ambient temperature and pressure (Andreozzi et al., 1999). They were initially developed for the sanitation of water and the destruction of environmental microorganisms, especially the pathogens (Liltved et al., 1995; Navarro et al., 2005; Rodriguez-Romoand and Yousef 2005; Shu and Chang 2005b; Moreau et al., 2008). It has been reported that the toxic contaminants and those which are not amenable to biological

[*] Corresponding Author, Hakimieh- Tehranpars, Shahid Abbaspour Blvd., PO Code 16589-53571, PO Box 16765-1719, Tehran, Iran. Email: g.badalians@yahoo.com.

treatment processes are mineralized and converted to carbon dioxide, water, short-chain organic acids, and inorganic ions, typically less toxic and amenable to biodegradation. Thus, the advanced oxidation processes occurrence results only in a small amount of the persistent chemical species (Adewuyi 2005a; Moreau et al., 2008). The advanced oxidation processes (AOPs) have the common characteristic of employing the high reactive hydroxyl radicals (strong oxidizing agent) for the treatment of wastewaters containing refractory organic compounds (Andreozzi et al., 1999), even though the sources of the radicals are different (Kim and Ihm 2011).

The reactions between hydroxyl radicals and organic species are controlled by the hydroxyl radicals' mass transfer to the organic species or the production of hydroxyl radicals. The kinetics of the related reactions seems to be a first order reaction with respect to the hydroxyl radicals' concentration and the contaminant (Esplugas et al., 2002, Parsons 2004). The hydroxyl radical, i.e an unstable and reactive agent must be generated continuously in situ through chemical, photochemical, or sonochemical reactions (Mahamuni and Adewuyi 2010).

The formation of the hydroxyl radicals by decomposition of the water was proposed and confirmed by several researchers (Weiss 1944; Makino et al., 1982; Mahamuni and Adewuyi 2010). A chain reaction (the reactions 1 to 3) occurs between the hydroxyl radical and the organic compound R as follows (Cano Quiroz et al., 2010):

$$RH + OH^0 \leftrightarrow R^0 + H_2O \tag{1}$$

$$R^o + O_2 \leftrightarrow ROO^0 \tag{2}$$

$$ROO^0 + RH \leftrightarrow ROOH + R^o \tag{3}$$

The advantages of the advanced oxidation processes can be employed to integrate biological treatment processes with an oxidative degradation of toxic or refractory substances entering or leaving the biological stage, e.g., Fenton-biological treatment processes (Pignatello et al., 2006; Wang et al., 2008; Badawy et al., 2009; Wang et al., 2009; Li et al., 2010; Xie et al., 2010; Elmolla and Chaudhuri 2011; Wang et al., 2012; Perez et al., 2013; Wang et al., 2014).

FENTON PROCESSES

Fenton and related reactions encompass reactions of peroxides, i.e., usually H_2O_2 with ferrous or ferric ions to form active oxygen species for oxidizing the present organic or inorganic compounds, i.e., Iron catalyzed decomposition of H_2O_2 for in situ generation of hydroxyl radicals for destroying toxic organic compounds in aqueous solutions. Various iron species, such as iron salts, i.e., in Fe^{2+} or Fe^{3+} form, iron oxides, e.g., Fe_2O_3 and Fe_3O_4, and zero-valent iron have been employed for chemical degradation, i.e., conventional Fenton process, photochemical degradation, i.e., photo-Fenton process, and electrochemical degradation, i.e., electro-Fenton process. Other elements, such as chromium, cerium, copper, cobalt, manganese, and ruthenium with multiple redox state can also decompose hydrogen peroxide into hydroxyl radical through Fenton-like processes, so-called Non-Ferrous Fenton. This can also achieved employing electron transfer mechanism in zero-valent Al/O_2 system (Bokare and Choi 2014). The related ferrous Fenton reactions are as follows (the reactions 4 to 7) (Andreozzi et al., 1999):

$$Fe^{2+} + H_2O_2 \rightarrow Fe^{3+} + OH^- + OH^0 \tag{4}$$

$$Fe^{3+} + H_2O_2 \leftrightarrow H^+ + FeOOH^{2+} \tag{5}$$

$$FeOOH^{2+} \rightarrow HO_2^0 + Fe^{2+} \tag{6}$$

$$Fe^{3+} + H_2O_2 \leftrightarrow Fe^{2+} + H^+ + HO_2^0 \tag{7}$$

However, one of the main disadvantageous issues related is the higher Fe^{2+}consumption rate than its reduction (Babuponnusami and Muthukumar 2012). The required acidic condition is also a limiting factor for the ferrous-based advanced oxidation processes (Bokare and Choi 2014). Other limitations and/or disadvantages of the Fenton-based advanced oxidation processes regarding wastewater treatment purposes are also reported (Ventura et al., 2002; Vinu et al., 2005; Ai et al., 2008; Cano Quiroz et al., 2010).

PHOTO-ASSISTED FENTON (PHOTO-FENTON) PROCESSES

The Fenton-Fenton like reagents degradation rate of the organic contaminants can be accelerated by means of UV light-irradiation at wavelength values higher than 300 nm (Kiwi et al., 1993, Pulgarin and Kiwi, 1996). The photolysis of Fe^{3+}complexes accordingly results in the Fe^{2+} regeneration as follows (the equation 8) (Andreozzi et al., 1999):

$$Fe^{3+} + OH^- \xrightarrow{h\nu} Fe^{2+} + OH^0 \tag{8}$$

In the recent years, several studies have been conducted regarding photo-Fenton process potentials, advantages, and disadvantages, especially solar photo-Fenton and related multistage combined reactors (Martin et al., 2009; Trovo et al., 2009; Michael et al., 2012; Silva et al., 2013a; Silva et al., 2013b; Ahmed and Chiron 2014; Miralles-Cuevas et al., 2014).

PHOTO FERRIOXALATE/H_2O_2 (FE^{3+}-CHELATE/H_2O_2)

An improvement of photo assisted Fenton processes is the UV-vis/Ferrioxalate/H_2O_2 system which has been proposed (Safarzadeh-Amiri et al., 1996, Ikehata and Gamal El-Din,

2005) as the more efficient process than the photo-Fenton for the abatement of organic contaminants corresponding with the following reactions 9 to 11:

$$\left[Fe^{3}(C_2O_4)_3\right]^{3-} + hv \rightarrow \left[Fe^{2}(C_2O_4)_3\right]^{2-} + \left[C_2O_4^{\ 0-}\right] \tag{9}$$

$$\left[C_2O_4^{\ 0-}\right] + \left[Fe^{3}(C_2O_4)_3\right]^{3-} \rightarrow \left[Fe^{2}(C_2O_4)_3\right]^{2-} + C_2O_4^{2-} + 2CO_2 \tag{10}$$

$$\left[C_2O_4^{\ 0-}\right] + O_2 \rightarrow O_2^{0-} + 2CO_2 \tag{11}$$

Irradiation of the ferrioxalate in an acidic solution generates carbon dioxide and ferrous ions Fe^{2+} free or the complexes with oxalate which in a combination with H_2O_2 provide a continuous source of Fenton's reagent. It has been reported that the energy required for treating the same volume of selected wastewater (toluene, xylenes, etc.) is about 20% of the energy required by photo-Fenton system (Safarzadeh-Amiri et al., 1997).

ACTIVATED PERSULFATE OXIDATION

As an alternative to the hydroxyl radical, the sulfate radical ($SO_4^{\bullet-}$) generated from persulfate ($S_2O_8^-$) has shown a high potential for degradation of persistant contaminants in water and wastewater, i.e., sulfate radical-based advanced oxidation process.

Several research results have been reported according to the activated persulfate oxidation and related combined methods as an advanced oxidation process with high removal efficiency for COD, color, NH_3-N, triclosan, microcystin-LR, acetic acid, phenol, azo dye acid orange 7, azo dye orange G, diphenylamine, trichloroethylene, BTEX, TCE, and volatile organic compounds (Huang et al., 2005; Liang et al., 2006; Liang et al., 2007; Liang et al., 2008, Liang et al., 2008; Criquet and Vel Leitner 2009; Li et al., 2009; Liang et al., 2009; Yang et al., 2009; Antoniou et al., 2010; Xu and Li 2010; Boukari et al., 2011; Criquet and Vel Leitner 2011; Deng and Ezyske 2011; Kusic et al., 2011; Lin et al., 2011; Nfodzo and Choi 2011; Yang et al., 2011; Abu Amr and Abdul Aziz 2012; Abu Amr et al., 2013).

SEMICONDUCTOR PHOTOCATALYSIS

Semiconductor photocatalysis has been used for the purification of air (Jacoby et al., 1996), water (Maness et al., 1999, Herrera Melián et al., 2000, Méndez-Hermida et al., 2007) and destruction of phytopathogens (Yao et al., 2007, Moreau et al., 2008).This process makes use of a semiconductormetal oxide as catalyst and of oxygen as oxidizing agent (Ollis and Al-Ekabi 1993, Andreozzi et al., 1999). In this regard, the anatase titanium dioxide (TiO_2) seems to have the most interesting attributes such as fast electron transfer to molecular oxygen, high photochemical reactivity, stability in aqueous systems and relatively low environmental toxicity (Linsebigler et al., 1995; Stafford et al., 1996; Gouvea et al., 2000; Daneshvar et al., 2004; Vinu and Madras 2008; Huang et al., 2009; Georgieva et al., 2012; Wu et al., 2012; Lu et al., 2012; Elizalde-Gonzalez et al., 2013; Liu et al., 2013; Tong et al., 2013; Gao et al., 2014; Larue et al., 2014; Li et al., 2014; Zhang et al., 2014; Zhang et al., 2014; Ahmad et al., 2015; Lee et al., 2015; Xu et al., 2015).

The use of semiconductor photocatalysis for the remediation of organic and inorganic contaminants has been the subject of several recent studies and reviews (Ollis 1985; Ollis 1986; Schiavello 1988; Ollis et al., 1991; Mills et al., 1993; Bahnemann et al., 1994; Hoffman et al., 1994; Hoffmann et al., 1995; Linsebigler et al., 1995; Mills et al., 1996; Vinodgopal et al., 1996; Fujishima et al., 2000; Yamazaki et al., 2001; Bhatkhande et al., 2002; Kamble et al., 2003; Pichat 2003; Shiraishi et al., 2003, Wong and Chu 2003; Kositzi et al., 2004; Pareek and Adesina 2004; Adewuyi, 2005b; Rumi et al., 2005; Haque et al., 2006;Munoz and Guieysse, 2006; Karunakaran et al., 2010; Yan et al., 2010; Li et al., 2012; Robertson et al., 2012).

The semiconductor photocatalysis process occurrence based on the absorption of a photon with energy equals to or bigger than the semiconductor's band-gapwhich that makes the electrons excite from the valence band to the conduction band possible. As a result, a photogenerated hole is created in the valence band (h_{VB}^{+}) and a photogenerated electron in the conduction band (e_{CB}^{-}) forming an electron-hole pair (the reaction 12) (Gouvea et al., 2000; Robertson et al., 2012):

$$photocatalyst \xrightarrow{hv} e_{CB}^{-} + h_{VB}^{+} \quad (12)$$

In semiconductor photocatalysis process hydroxyl radicals are the main reactive species (Daneshvar et al., 2004) which are formed when photo-generated holes are scavenged by surface hydroxyl groups to generate hydroxyl radical (the reaction 13) (Robertson et al., 2012).

$$h_{VB}^{+} + OH_{s} \rightarrow OH_{s}^{0} \quad (13)$$

The suffix "s" in $OH_s^{\bullet}$ indicates the formation of surface groups and can thus be interpreted as surface trapped holes.

These reactions are of great importance in oxidative degradation processes due to the high concentration of H_2O and HO^- adsorbed on the particle surface (Andreozzi et al., 1999).

The formation of •OH can follow the reductive pathway of photogenerated electrons which react with a proper acceptor; this is in environmental photocatalysis normally assumed to be O_2 (the reaction 15). (Robertson et al., 2012):

$$e_{CB}^{-} + O_2 \rightarrow O_2^{0-} \rightarrow H_2O_2 \rightarrow OH^0 + OH^- \quad (14)$$

The initially formed superoxide radical results finally in H_2O_2 which is then further reduced yielding hydroxyl radicals.

Substrates can also be directly oxidized by electron transfer (the reaction 15) (Andreozzi et al., 1999; Daneshvar et al., 2004; Robertson et al., 2012):

$$h_{VB}^{+} + contam._{s} \rightarrow contam._{s}^{0+} \quad (15)$$

The oxidation process using hydroxyl radical is dominant in a system with a weak contaminant adsorption while direct oxidation (the reaction 15) is associated with a system where the contaminant is adsorbed strongly at the photocatalyst surface establishing a good electronic contact (Robertson et al., 2012).

A significant part of electron-hole pairs recombine thus reducing the quantum yield (Andreozzi et al., 1999; Robertson et al., 2012). One way to overcome this problem is to add a proper dosage of the hydrogen peroxide according to the kind and the concentration of the contaminant (Bauer et al., 1999). According to the reaction 16, a low concentration of the hydrogen peroxide inhibits the electron–hole recombination. Since the hydrogen peroxide is a better electron acceptor than the molecular oxygen, it could act as an alternative electron acceptor to oxygen (Khodja et al., 2001; Daneshvar et al., 2003). Hydrogen peroxide can also be photolyzed resulting in the hydroxyl radicals' production directly by UV/H_2O_2 process (the reaction 17) (Stefan and Bolton 1999; Cater et al., 2000; Daneshvar et al., 2004).

$$e_{CB}^{-} + H_2O_2 \rightarrow OH^{-} + OH^{0} \tag{16}$$

$$H_2O_2 + hv \rightarrow 2OH^{0} \tag{17}$$

However the hydrogen peroxide at high dosage is a powerful •OH scavenger (the reaction 18) (Daneshvar et al., 2003; Daneshvar et al., 2004):

$$H_2O_2 + OH^{0} \rightarrow HO_2^{0} + H_2O \tag{18}$$

Zinc oxide appears to be a suitable alternative to TiO_2 since its photo degradation mechanism has been proven to be similar to TiO_2 (Serpone et al., 1995; Stafford et al., 1996; Villaseñor and Mansilla 1996; Yeber et al., 1999; Gouvea et al., 2000; Dindar and Içli 2001; Khodja et al., 2001; Marc et al., 2001; Pirkanniemi and Sillanpaeae 2002; Daneshvar et al., 2004; Daneshvar et al., 2004). The observed higher efficiency for the ZnO semiconductor can be explained on account of its higher adsorption capacity (Gouvea et al., 2000).Modified (doped) TiO_2 have also been developed to increase adsorption capacity of TiO_2 (Andreozzi et al., 1999; Gouvea et al., 2000).

Many studies also report a significant enhancement of the photochemical activity by coupling semiconductors such as ZnO/TiO_2 (Hotchandani and Kamat 1992; Jardim et al., 1993; Ranjit and Viswanathan 1997; Lin and Yu 1998; Gouvea et al., 2000). This enhancement is the result of an efficient modifying electronic properties, i.e., the electron transfer from the conduction band of the ZnO to the conduction band of TiO_2 under illumination, and conversely, the hole transfer from the valence band of TiO_2 to the valence band of ZnO (Serpone et al., 1995) give rise to a decrease of the pairs recombination rate (Marci et al., 2001). Some other researchers have Supported highly photoactive species on particles resistant to mixing in water to minimize the problem of separating the photocatalyst from the reacting aqueous suspension, such as TiO_2 (anatase) on Al_2O_3, TiO_2 (anatase) on TiO_2 (rutile), and ZnO on sand (Marci et al., 2001).

OZONATION (PH>8)

Over last decades, ozone has been used as an oxidant for water and wastewater treatment (Horváth et al., 1985; Catalkaya and Kargi 2007; Deborde et al., 2008; Garoma and Matsumoto 2009; Ko et al., 2009; Somensi et al., 2010; Antoniou et al., 2013; Ibanez et al., 2013; Jagadevan et al., 2013; Margot et al., 2013; Umar et al., 2013; Barndok et al., 2014). It has been most widely employed for disinfecting water (Lezcano et al., 2000, Lezcano et al., 2001, Mysore et al., 2004, Moreau et al., 2008).

The ozonation process at a pH value equals to more than 8 is considered as an advanced oxidation process due to the ozone molecules decomposition into hydroxyl radicals and the occurred reaction between the hydroxyl radicals and organics afterwards. The hydroxyl radicals produced by advanced oxidation processes have a higher oxidation potential, i.e., equals to 2.8 V than molecular ozone. Therefore, it attacks the organic and inorganic molecules non-selectively with a high reaction rate. Consequently, the best results were achieved in the alkaline medium (Camel and Bermond, 1998; Andreozzi et al., 1999; Elovitz et al., 2000; Park et al., 2001; Flyunt et al., 2003; Ikehata and Gamal El-Din 2004; Ikehata and Gamal El-Din, 2005; Peternel et al., 2006; Noethe et al., 2009; Gao et al., 2014).

Ozone technologies occurrence has been investigated in various areas, i.e., pesticides degradation (Ikehata and Gamal El-Din 2005), micro contaminant degradation (Huber et al., 2004; Suarez et al., 2007; Dewitte et al., 2008; Schmidt and Brauch 2008; Andrzejewski et al., 2008; Oya et al., 2008; Zhao et al., 2008; Dodd et al., 2009; Hollender et al., 2009,), recycling marine aquaria, electroplating wastes, electronic chip manufacture, textiles, and petroleum refineries, paper pulp bleaching, rubber additive wastewaters, landfill leachates, and detergents in municipal wastewaters (Rice 1996).

Ozone chemistry in aqueous alkaline solutions has been extensively and successfully studied in the attempt of giving a chemical explanation to the short life time of ozone in alkaline solutions (Hoigné 1998, Andreozzi et al., 1999). The related results showed that the ozone decomposition in aqueous solution develops through the formation of hydroxyl radicals (the reactions 19 to 25). OH^- ion has in the reaction mechanism the role of an initiator:

$$HO^- + O_3 \rightarrow O_2 + HO_2^- \xleftrightarrow{H^+} H_2O_2 \tag{19}$$

$$O_3 + HO_2^- \xleftrightarrow{+} HO_2^0 + O_3^{0-} \tag{20}$$

$$HO_2^0 \leftrightarrow O_2^{0-} + H^+ \tag{21}$$

$$O_2^0 + O_3 \rightarrow O_3 + O_3^{0-} \tag{22}$$

$$O_3^{0-} + H^+ \rightarrow HO_3^0 \tag{23}$$

$$HO_3^0 \rightarrow HO^0 + O_2 \tag{24}$$

$$HO^0 + O_3 \rightarrow HO_2^0 + O_2 \qquad (25)$$

The simplified mechanism can be presented as follows (the reaction 26). (Tomiyasu et al., 1985, Peternel et al., 2006):

$$O_3 + H_2O \xrightarrow{OH^-} HO^0 + O_2 + HO_2^0 \qquad (26)$$

The influence of the pH value in ozonation process is evident. The increase of the pH value results into higher rates of the hydroxyl radicals' production (Glaze and Kang 1989, Andreozzi et al., 1999). The combination of ozonation process with other technologies, such as UV, ultrasound, electrocoagulation, and flotation allows an increase of the contaminants removal yield (Shu and Huang 1995; Mare et al., 1999; Shu and Chang 2005a; Abramov et al., 2006; Saritha et al., 2007; Ikehata et al., 2008; Beneventi et al., 2009; Rosal et al., 2009; Cano Quiroz et al., 2010; Kusvuran et al., 2010; Hernández-Ortega et al., 2010; Biń and Sobera-Madej, 2012; Wang et al., 2012). The related processes investigated are as follows:

O_3/H_2O_2 (Peroxone Process)

Ozonation process is very effective in the quality improvement of municipal wastewater effluent by eliminating the organic microcontaminants that react rapidly with ozone. However, there are ozone-persistent microcontaminants such as organic phosphates and others. They are decomposed upon ozonation to some extent due to hydroxyl radicals produced in the ozone reaction with organics in wastewater. The reaction of ozone with the anion of hydrogen peroxide, i.e., peroxone process gives rise to the hydroxyl radicals (Forni et al., 1982; Staehelin and Hoigne 1982; Glaze and Kang 1989; Spanggord et al., 2000; Rosal et al., 2008; Noethe et al., 2009; Rosal et al., 2009; Merenyi et al., 2010; Pocostales et al., 2010; Cyris et al., 2013; Fischbacher et al., 2013; Lee et al., 2013).

UV/H_2O_2 (H_2O_2 Photolysis)

The UV/H_2O_2 technique includes the photolysis of hydrogen peroxide (H_2O_2) to produce hydroxyl radicals ($^{\bullet}OH$) by ultraviolet (UV) radiation (254 nm), which are very effective in the oxidative degradation of most organic contaminants (Colonna et al., 1999; Ince 1999; Stefan and Bolton 1999; Cater et al., 2000; Garcia Einschlag et al., 2002; Beltran, 2003; Sharpless and Linden 2003; Aleboyeh et al., 2005; Shu and Chang 2005a; Dolinova et al., 2006; Peternel et al., 2006; Sarathy and Mohseni 2007; Dodd et al., 2010; Afzal et al., 2012; Kelly and Arnold 2012; Farre et al., 2012; Forsyth et al., 2013; Keen and Linden 2013; Zheng and Arnold 2013; Chu et al., 2014; He et al., 2014; Li et al., 2014).

The photolysis process rate of the aqueous hydrogen peroxide is pH dependent (Legrini et al., 1993). This may be primarily due to the high molar absorption coefficient of the peroxide anion HO_2^- (Andreozzi et al., 1999).

Several studies have been conducted during the last decade for improving the understanding of the UV/H_2O_2 advanced oxidation process and its kinetic modeling (Li et al., 2007; Rosenfeldt and Linden 2007; Popov et al., 2010; Santoro et al., 2010; Yao et al., 2013).

O_3/UV (PHOTOLYTIC OZONATION)

The first research results regarding photolytic ozonation process showed that O_3/UV enhances the oxidation of complexes cyanides, chlorinated solvents, pesticides, organics and miscellaneous group parameters, such as COD and BOD (Garrison et al., 1975, Prengle et al., 1975, Prengle Jr et al., 1976, Prengle Jr, 1977). Subsequently, further studies have been conducted for extending the related applications and improving the process efficiency (Fochtman and Huff, 1975; Garrison et al., 1975; Lee et al., 1977; Kuo et al., 1978; Glaze 1980; Glaze et al., 1982; Peyton et al., 1982; Glaze 1984; Peyton and Glaze, 1987; Ku et al., 2000; Benitez et al., 2002; Irmak et al., 2005; Chen et al., 2007; Shang et al., 2007; Wang et al., 2009; Rao and Chu 2010; Xu et al., 2010).

Ozone dissolved in water decomposes in the presence of UV light to H_2O_2 according to the following reaction (Beltran, 2003, Peternel et al., 2006):

$$O_3 + H_2O + hv \rightarrow H_2O_2 + O_2 \tag{27}$$

And then, it proceeds to the formation of hydroxyl radicals, as shown in reaction 17 (Andreozzi et al., 1999). Consequently, higher hydroxyl radicals' generation in the bulk is obtained than in the case of only indirect ozonation (Peternel et al., 2006).

The O_3/UV process has the chemical behavior similar to both O_3/H_2O_2 and H_2O_2/UV systems (Legrini et al., 1993; Andreozzi et al., 1999).

O_3/H_2O_2/UV

Research results have shown that the O_3/ H_2O_2/UV process represents higher degradation rate and occurs faster than H_2O_2/UV and O_3/UV processes (Esplugas et al., 2002; Azbar et al., 2004; Peternel et al., 2006; Popiel et al., 2008). The reason might be the higher generation rate of the hydroxyl radicals compared to the H_2O_2/UV and O_3/UV processes.The mechanism of the O_3/H_2O_2/UV process (the reaction 28) is the sum of the reactions 17and 27:

$$O_3 + H_2O_2 + H_2O \xrightarrow{hv} 4OH^0 + O_2 \tag{28}$$

The UV-based processes using ozone as an additional oxidant possesses a constant mineralization rate during the treatment occurrence time due to the constant ozone supply to the reactor resulting in a constant generation of hydroxyl radicals, as shown in the reactions 17 and 27 (Peternel et al., 2006).

References

Abramov, V., Abramov, O., Gekhman, A., Kuznetsov, V. & Price, G. (2006) Ultrasonic intensification of ozone and electrochemical destruction of 1, 3-dinitrobenzene and 2, 4-dinitrotoluene, *Ultrasonics sonochemistry* 13, Elsevier, 303-307.

Abu Amr, S.S., Abdul Aziz, H. (2012) New treatment of stabilized leachate by ozone/Fenton in the advanced oxidation process, *Waste Management* 32(9), Elsevier, 1693-1698.

Abu Amr, S.S., Abdul Aziz, H., Adlan, M.N. & Bashir, M.J.K. (2013) Pretreatment of stabilized leachate using ozone/persulfate oxidation process, *Chemical Engineering Journal* 221, Elsevier, 492-499.

Adewuyi, Y.G. (2005a) Sonochemistry in environmental remediation. 1. Combinative and hybrid sonophotochemical oxidation processes for the treatment of pollutants in water, Environmental Science & Technology 39, *American Chemical Society* (ACS), 3409-3420.

Adewuyi, Y.G. (2005b) Sonochemistry in environmental remediation. 2. Heterogeneous sonophotocatalytic oxidation processes for the treatment of pollutants in water, Environmental Science & Technology 39, *American Chemical Society* (ACS), 8557-8570.

Afzal, A., Drzewicz, P., Perez-Estrada, A.L., Chen, Y., Martin, J.W. & Gamal El-Din, M. (2012) Effect of Molecular Structure on the Relative Reactivity of Naphthenic Acids in the UV/H_2O_2 Advanced Oxidation Process, Environmental Science & Technology 46(19), *American Chemical Society (ACS),* 10727-10734.

Ahmed, M.M., Chiron, S. (2014) Solar photo-Fenton like using persulphate for carbamazepine removal from domestic wastewater, Water Research 48, *the International Water Association (IWA),* Elsevier, 229-236.

Ahmad, R., Mohsin, M., Ahmad, T. & Sardar, M. (2015) Alpha amylase assisted synthesis of TiO_2 nanoparticles: Structural characterization and application as antibacterial agents, *Journal of Hazardous Materials* 283, Elsevier, 171-177.

Ai, Z., Xiao, H., Mei, T., Liu, J., Zhang, L., Deng, K. & Qiu, J. (2008) Electro-Fenton degradation of rhodamine B based on a composite cathode of Cu2O nanocubes and carbon nanotubes, The Journal of Physical Chemistry C 112, *American Chemical Society (ACS),* 11929-11935.

Aleboyeh, A., Moussa, Y. & Aleboyeh, H. (2005) The effect of operational parameters on UV/H2O2 decolourisation of Acid Blue 74, *Dyes and Pigments* 66, Elsevier, 129-134.

Andreozzi, R., Caprio, V., Insola, A. & Marotta, R. (1999) Advanced oxidation processes (AOP) for water purification and recovery, *Catalysis Today* 53, Elsevier, 51-59.

Andrzejewski, P., Kasprzyk-Hordern, B. & Nawrocki, J. (2008) < i> N</i>-nitrosodimethylamine (NDMA) formation during ozonation of dimethylamine-containing waters, *Water Research 42, the International Water Association* (IWA), Elsevier, 863-870.

Antoniou, M.G., de la Cruz, A.A. & Dionysiou, D.D. (2010) Degradation of microcystin-LR using sulfate radicals generated through photolysis, thermolysis and e^- transfer mechanisms, *Applied Catalysis B: Environmental 96*(3-4), Elsevier, 290-298.

Antoniou, M.G., Hey, G., Rodriguez Vega, S., Spiliotopoulo, A., Fick, J., Tysklind, M., la Cour Jansen, J. & Andersen, H.R. (2013) Required ozone doses for removing pharmaceuticals from wastewater effluents, *Science of The Total Environment* 456-457, Elsevier, 42-49.

Azbar, N., Yonar, T. & Kestioglu (2004) Comparison of various advanced oxidation processes and chemical treatment methods for COD and color removal from a polyester and acetate fiber dyeing effluent, *Chemosphere* 55(1), Elsevier, 35-43.

Babuponnusami, A., Muthukumar, K. (2012) Advanced oxidation of phenol: a comparison between Fenton, electro-Fenton, sono-electro-Fenton and photo-electro-Fenton processes, *Chemical Engineering Journal* 183, Elsevier, 1-9.

Badawy, M.I., Wahab, R.A. & El-Kalliny, A.S. (2009) Fenton-biological treatment processes for the removal of some pharmaceuticals from industrial wastewater, *Journal of Hazardous Materials* 167(1-3), Elsevier, 567-574.

Bahnemann, D., Cunningham, J., Fox, M., Pelizzetti, E., Pichat, P. & Serpone, N. (1994) Photocatalytic treatment of waters, *Aquatic and surface photochemistry,* edited by G.R. Helz, R.G. Zepp and D.G. Crosby, Lewis Publishers, 261-316.

Barndok, H., Cortijo, L., Hermosilla, D., Negro, C. & Blanco, A. (2014) Removal of 1,4-dioxane from industrial wastewaters: Routes of decomposition under different operational conditions to determine the ozone oxidation capacity, *Journal of Hazardous Materials* 280, Elsevier, 340-347.

Bauer, R., Waldner, G., Fallmann, H., Hager, S., Klare, M., Krutzler, T., Malato, S. & Maletzky, P. (1999) The photo-Fenton reaction and the TiO2/UV process for waste water treatment- novel developments, *Catalysis Today* 53, Elsevier, 131-144.

Beltran, F.J. (2003) Ozone-UV radiation-hydrogen peroxide oxidation technologies, *Environmental Science and Pollution Control Series,* 1-76.

Beneventi, D., Almeida, F., Marlin, N., Curtil, D., Salgueiro, L. & Aurousseau, M. (2009) Hydrodynamics and recovered papers deinking in an ozone flotation column, *Chemical Engineering and Processing: Process Intensification* 48, Elsevier, 1517-1526.

Benitez, F.J., Acero, J.L. & Real, F.J. (2002) Degradation of carbofuran by using ozone, Uv radiation and advanced oxidation processes, *Journal of Hazardous Materials* 89(1), Elsevier, 51-65.

Bhatkhande, D.S., Pangarkar, V.G. & Beenackers, A.A. (2002) Photocatalytic degradation for environmental applications - a review, Journal of Chemical Technology and Biotechnology 77, *Society of Chemical Industry (SCI),* Wiley, 102-116.

Bianco, B., de Michelis, I. & Veglio, F. (2011) Fenton treatment of complex industrial wastewater : Optimization of process conditions by surface response method, *Journal of Hazardous Materials* 186(2-3), Elsevier, 1733-1738.

Bin, A.K., Sobera-Madej, S. (2012) Comparison of the advanced oxidation processes (UV, UV/H2O2 and O3) for the removal of antibiotic substances during wastewater treatment, *Ozone: Science & Engineering 34, the International Ozone Association,* Taylor & Francis, 136-139.

Bokare, A.D., Choi, W. (2014) Review of iron-free Fenton-like systems for activating H_2O_2 in advanced oxidation processes, *Journal of Hazardous Materials* 275, Elsevier, 121-135.

Boukari, S.O.B., Pellizzari, F. & Vel Leitner, N.K. (2011) Influence of persulfate ions on the removal of phenol in aqueous solution using electron beam irradiation, *Journal of Hazardous Materials* 185(2-3), Elsevier, 844-851.

Buxton, G.V., Greenstock, C.L., Helman, W.P. & Ross, A.B. (1988) Critical review of rate constants for reactions of hydrated electrons, hydrogen atoms and hydroxyl radicals in aqueous solution, Journal of physical and chemical reference data 17, *American Institute of Physics* (AIP), 513-886.

Camel, V., Bermond, A. (1998) The use of ozone and associated oxidation processes in drinking water treatment, *Water Research 32, the International Water Association* (IWA), Elsevier, 3208-3222.

Cano Quiroz, A., Barrera-Diaz, C., Roa-Morales, G., Balderas HernAandez, P., Romero, R. & Natividad, R. (2010) Wastewater ozonation catalyzed by iron, *Industrial & Engineering Chemistry Research 50, American Chemical Society,* 2488-2494.

Catalkaya, E.C., Kargi, F. (2007) Color, TOC and AOX removals from pulp mill effluent by advanced oxidation processes: A comparative study, *Journal of Hazardous Materials* 139(2), Elsevier, 244-253.

Cater, S.R., Stefan, M.I., Bolton, J.R. & Safarzadeh Amiri, A. (2000) UV/H_2O_2 treatment of methyl tert-butyl ether in contaminated waters, *Environmental Science & Technology 34(4), American Chemical Society,* 659-662.

Chen, W.S., Juan, C.N. & Wei, K.M. (2007) Decomposition of dinitrotoluene isomers and 2,4,6-trinitrotoluene in spent acid from toluene nitration process by ozonation and photo-ozonation, *Journal of Hazardous Materials* 147(1-2), Elsevier, 97-104.

Chu, W., Gao, N., Yin, D., Krasner, S.W. & Mitch, W.A. (2014) Impact of UV/H_2O_2 Pre-Oxidation on the Formation of Haloacetamides and Other Nitrogenous Disinfection Byproducts during Chlorination, Environmental Science & Technology 48(20), *American Chemical Society,* 12190-12198.

Colonna, G.M., Caronna, T. & Marcandalli, B. (1999) Oxidative degradation of dyes by ultraviolet radiation in the presence of hydrogen peroxide, *Dyes and Pigments* 41, Elsevier, 211-220.

Criquet, J., Vel Leitner, N.K. (2009) Degradation of acetic acid with sulfate radical generated by persulfate ions phytolysis, *Chemosphere* 77(2), Elsevier, 194-200.

Criquet, J., Vel Leitner, N.K. (2011) Electron beam irradiation of aqueous solution of persulfate ions, *Chemical Engineering Journal* 169(1-3), Elsevier, 258-262.

Cyris, M., Knolle, W., Richard, J., Dopp, E., von Sonntag, C. & Schmidt, T.C. (2013) Reaction of Gadolinium Chelates with Ozone and Hydroxyl Radicals, *Environmental Science & Technology 47(17), American Chemical Society* (ACS), 9942-9949.

Daneshvar, N., Salari, D. & Khataee, A. (2003) Photocatalytic degradation of azo dye acid red 14 in water: investigation of the effect of operational parameters, *Journal of Photochemistry and Photobiology A: Chemistry* 157, 111-116.

Daneshvar, N., Salari, D. & Khataee, A. (2004) Photocatalytic degradation of azo dye acid red 14 in water on ZnO as an alternative catalyst to TiO_2, *Journal of Photochemistry and Photobiology A: Chemistry* 162, Elsevier, 317-322.

Debellefontaine, H., Chakchouk, M., Foussard, J., Tissot, D. & Striolo, P. (1996) Treatment of organic aqueous wastes: Wet air oxidation and wet peroxide oxidation< sup>®</sup>, *Environmental Pollution* 92, Elsevier, 155-164.

Deborde, M., Rabouan, S., Mazellier, P., Duguet, J.P. & Legube, B. (2008) Oxidation of bisphenol A by ozone in aqueous soluton, *Water Research* 42(16), the International Water Association (IWA), Elsevier, 4299-4308.

Deng, Y., Ezyske, C.M. (2011) Sulfate radical-advanced oxidation processes (SR-AOP) for simultaneous removal of refractory organic contaminants and ammonia in landfill leachate, *Water Research* 45(18), the International Water Association (IWA), Elsevier, 6189-6194.

Dewitte, B., Dewulf, J., Demeestere, K., van de Vyvere, V., de Wispelaere, P. & van Langenhove, H. (2008) Ozonation of ciprofloxacin in water: HRMS identification of reaction products and pathways, *Environmental Science & Technology* 42, American Chemical Society (ACS), 4889-4895.

Dindar, B., Icli, S. (2001) Unusual photoreactivity of zinc oxide irradiated by concentrated sunlight, *Journal of Photochemistry and Photobiology A: Chemistry* 140, Elsevier, 263-268.

Dodd, M.C., Kohler, H.P.E. & von Gunten, U. (2009) Oxidation of antibacterial compounds by ozone and hydroxyl radical: elimination of biological activity during aqueous ozonation processes, *Environmental Science & Technology* 43, American Chemical Society (ACS), 2498-2504.

Dodd, M.C., Rentsch, D., Singer, H.P., Kohler, H.P.E. & von Gunten, U. (2010) Transformation of β-Lactam Antibacterial Agents during Aqueous Ozonation: Reaction Pathways and Quantative Bioassay of Biologically-Active Oxidation Products, *Environmental Science & Technology* 44(15), American Chemical Society, 5940-5948.

Dolinova, J., Ruzicka, R., Kurkova, R., Klanova, J. & Klan, P. (2006) Oxidation of Aromatic and Aliphatic Hydrocarbons by OH Radicals Photochemically Generated from H_2O_2 in Ice, Environmental Science & Technology 40(24), *American Chemical Society,* 7668-7674.

Eisenhauer, H.R. (1964) Oxidation of phenolic wastes, *Journal of Water Pollution Control Federation,* 1116-1128.

Elizalde-Gonzalez, M.P., Garcia-Diaz, E. & Sabinas-Hernandez, S.A. (2013) Novel preparation of carbon-TiO_2 composites, *Journal of Hazardous Materials* 263(1), Elsevier, 73-83.

Elmolla, E.S., Chaudhuri, M. (2011) Combined photo-Fenton-SBR process for antibiotic wastewater treatment, *Journal of Hazardous Materials* 192(3), Elsevier, 1418-1426.

Elovitz, M.S., von Gunten, U. & Kaiser, H.P. (2000) Hydroxyl radical/ozone ratios during ozonation processes. II. The effect of temperature, pH, alkalinity, and DOM properties, *Ozone: Science & Engineering* 22, the International Ozone Association, Taylor & Francis, 123-150.

Esplugas, S., Gimenez, J., Contreras, S., Pascual, E. & Rodriguez, M. (2002) Comparison of different advanced oxidation processes for phenol degradation, *Water Research* 36(4), the International Water Association (IWA), Elsevier, 1034-1042.

Farre, M.J., Radjenovic, J. & Gernjak, W. (2012) Assessment of Degradation Byproducts and NDMA Formation Potential during UV and UV/H_2O_2 Treatment of Doxylamine in the Presence of Monochloramine, *Environmental Science & Technology* 46(23), American Chemical Society, 12904-12912.

Fischbacher, A., von Sonntag, J., von Sonntag, C. & Schmidt, T.C. (2013) The $^{\bullet}OH$ Radical Yield in the H_2O_2+O_3 (Peroxone) Reaction, *Environmental Science & Technology* 47(17), American Chemical Society (ACS), 9959-9964.

Flyunt, R., Leitzke, A., Mark, G., Mvula, E., Reisz, E., Schick, R. & von Sonntag, C. (2003) Determination of OH, O_2-, and hydroperoxide yields in ozone reactions in aqueous solution, *The Journal of Physical Chemistry* B 107, American Chemical Society (ACS), 7242-7253.

Fochtman, E., Huff, J. (1975) Ozone-Ultraviolet Light Treatment of TNT Wastewaters, *Water Pollution* 13, 211-223.

Forni, L., Bahnemann, D. & Hart, E.J. (1982) Mechanism of the hydroxide ion-initiated decomposition of ozone in aqueous solution, *The Journal of Physical Chemistry* 86, American Chemical Society (ACS), 255-259.

Forsyth, J.E., Zhou, P., Mao, Q., Asato, S.S., Meschke, J.S. & Dodd, M.C. (2013) Enhanced Inactivation of Bacillus subtilis Spores during Solar Photolysis of Free Available Chlorine, *Environmental Science & Technology* 47(22), American Chemical Society (ACS), 12976-12984.

Fujishima, A., Rao, T.N. & Tryk, D.A. (2000) Titanium dioxide photocatalysis, Journal of Photochemistry and Photobiology C: Photochemistry Reviews 1, Elsevier, 1-21.

Gao, P., Li, A., Sun, D.D. & Ng, W.J. (2014) Effects of various TiO_2 nanostructures and graphene oxide on photocatalytic activity of TiO_2, *Journal of Hazardous Materials* 279, Elsevier, 96-104.

Gao, S., Zhao, Z., Xu, Y., Tian, J., Qi, H., Lin, W. & Cui, F. (2014) Oxidation of sulfamethoxazole (SMX) by chlorine, ozone and permanganate – A comparative study, *Journal of Hazardous Materials* 274, Elsevier, 258-269.

Garoma, T., Matsumoto, S. (2009) Ozonation of aqueous solution containing bisphenol A: Effect of operational parameters, *Journal of Hazardous Materials* 167(1-3), Elsevier, 1185-1191.

Garrison, R., Mauk, C. & Prengle Jr, H. (1975) Advanced ozone oxidation system for complexed cyanides, *Ozone for Water and Waste Water Treatment, International Ozone Association, Cleveland*, OH, 551-577.

Garcia Einschlag, F.S., Lopez, J., Carlos, L., Capparelli, A.L., Braun, A.M. & Oliveros, E. (2002) Evalution of the Efficiency of Photodegradation of Nitroaromatics Applying the UV/H_2O_2 Technique, *Environmental Science & Technology* 36(18), American Chemical Society (ACS), 3936-3944.

Georgieva, J., Valova, E., Armyanov, S., Philippidis, N., Poulios, I. & Sotiropoulos, S. (2012) Bi-component semiconductor oxide photoanodes for the photoelectrocatalytic oxidation of organic solutes and vapours: A short review with emphasis to TiO_2-WO_3 photoanodes, *Journal of Hazardous Materials 211-212,* Elsevier, 30-46.

Glaze, W. (1980) Oxidation of water supply refractory species by ozone with ultraviolet radiation, *Environmental protection technology series,* USA.

Glaze, W.H., Kang, J.W. (1989) Advanced oxidation processes. Description of a kinetic model for the oxidation of hazardous materials in aqueous media with ozone and hydrogen peroxide in a semibatch reactor, *Industrial & Engineering Chemistry Research* 28, American Chemical Society (ACS), 1573-1580.

Glaze, W.H., Peyton, G.R., Lin, S., Huang, R. & Burleson, J.L. (1982) Destruction of pollutants in water with ozone in combination with ultraviolet radiation. II. Natural trihalomethane precursors, *Environmental Science & Technology 16, American Chemical Society (*ACS), 454-458.

Gouvea, C. A., Wypych, F., Moraes, S. G., Duran, N., Nagata, N. & Peralta-Zamora, P. (2000) Semiconductor-assisted photocatalytic degradation of reactive dyes in aqueous solution, *Chemosphere* 40, Elsevier, 433-440.

Haque, M.M., Muneer, M. & Bahnemann, D.W. (2006) Semiconductor-Mediated Photocatalyzed Degradation of a herbicide Derivate, Chlorotoluron, in Aqueous

Suspensions, Environmental Science & Technology 40(15), *American Chemical Society* (ACS), 4765-4770.

He, X., Zhang, G., de la Cruz, A.A., O'Shea, K.E. & Dionysiou, D.D. (2014) Degradation Mechanism of Cyanobacterial Toxin Cylindrospermopsin by hydroxyl Radicals in Homogeneous UV/ H_2O_2 Process, *Environmental Science & Technology* 48(8), American Chemical Society (ACS), 4495-4504.

Hernandez-Ortega, M., Ponziak, T., Barrera-Diaz, C., Rodrigo, M., Roa-Morales, G. & Bilyeu, B. (2010) Use of a combined electrocoagulation–ozone process as a pre-treatment for industrial wastewater, *Desalination* 250, Elsevier, 144-149.

Herrera Melian, J., Dona Rodriguez, J., Viera Suarez, A., Tello Rendon, E., Valdes Do Campo, C., Arana, J. & Perez Pena, J. (2000) The photocatalytic disinfection of urban waste waters, *Chemosphere* 41, Elsevier, 323-327.

Hoffmann, A.J., Carraway, E.R. & Hoffmann, M.R. (1994) Photocatalytic production of H_2O_2 and organic peroxides on quantum-sized semiconductor colloids, *Environmental Science &* Technology 28, American Chemical Society (ACS), 776-785.

Hoffmann, M.R., Martin, S.T., Choi, W. & Bahnemann, D.W. (1995) Environmental applications of semiconductor photocatalysis, *Chemical Reviews* 95, American Chemical Society (ACS), 69-96.

Hoigne, J. (1998) Chemistry of aqueous ozone and transformation of pollutants by ozonation and advanced oxidation processes, *Quality and Treatment of Drinking Water* II, Springer.

Hollender, J., Zimmermann, S.G., Koepke, S., Krauss, M., McArdell, C.S., Ort, C., Singer, H., Von Gunten, U. & Siegrist, H. (2009) Elimination of organic micropollutants in a municipal wastewater treatment plant upgraded with a full-scale post-ozonation followed by sand filtration, *Environmental Science & Technology* 43, American Chemical Society (ACS), 7862-7869.

Hotchandani, S., Kamat, P.V. (1992) Charge-transfer processes in coupled semiconductor systems. Photochemistry and photoelectrochemistry of the colloidal cadmium sulfide-zinc oxide system, *The Journal of Physical Chemistry* 96, American Chemical Society (ACS), 6834-6839.

Huang, H., Li, D., Lin, Q., Zhang, W., Shao, Y., Chen, Y., Sun, M. & Fu, X. (2009) Efficient Degradation of Benzene over $LaVO_4/TiO_2$ Nanocrystalline Heterojunction Photocatalyst under Visible Light Irradiation, *Environmental Science & Technology* 43(11), American Chemical Society (ACS), 4164-4168.

Huang, K.C., Zhao, Z., Hoag, G.E., Dahmani, A. & Block, P.A. (2005) Degradation of volatile organic compounds with thermaly activated persulfate oxidation, *Chemosphere* 61(4), Elsevier, 551-560.

Huber, M.M., Ternes, T.A. & Von Gunten, U. (2004) Removal of estrogenic activity and formation of oxidation products during ozonation of 17α-ethinylestradiol, *Environmental Science & Technology* 38, American Chemical Society (ACS), 5177-5186.

Ibanez, M., Garcia-Lor, E., Bijlsma, L., Morales, E., Pastor, L. & Hernandez, F. (2013) Removal of emerging contaminants in sewage water subjected to advanced oxidation with ozone, *Journal of Hazardous Materials* 260, Elsevier, 389-398.

Ikehata, K., Gamal El-Din, M.G. (2004) Degradation of recalcitrant surfactants in wastewater by ozonation and advanced oxidation processes: a review, *Ozone: Science & Engineering* 26, the International Ozone Association, Taylor & Francis, 327-343.

Ikehata, K., Gamal El-Din, M.G. (2005) Aqueous pesticide degradation by ozonation and ozone-based advanced oxidation processes: a review (Part I), *Ozone: Science and Engineering* 27, the International Ozone Association, Taylor & Francis, 83-114.

Ikehata, K., Gamal El-Din, M. & Snyder, S.A. (2008) Ozonation and advanced oxidation treatment of emerging organic pollutants in water and wastewater, *Ozone: Science and Engineering 30, the International Ozone Association,* Taylor & Francis, 21-26.

Ince, N.H. (1999) "Critical" effect of hydrogen peroxide in photochemical dye degradation, *Water Research* 33, the International Water Association (IWA), Elsevier, 1080-1084.

Irmak, S., Erbatur, O. & Akgerman, A. (2005) Degradation of 17β-estradiol and bisphenol A in aqueous medium by using ozone and ozone/UV techniques, *Journal of Hazardous Materials* 126(1-3), Elsevier, 54-62.

Jacoby, W.A., Blake, D.M., Penned, J.A., Boulter, J.E., Vargo, L.M., George, M.C. & Dolberg, S. K. (1996) Heterogeneous photocatalysis for control of volatile organic compounds in indoor air, *Journal of the Air & Waste Management Association* 46, 891-898.

Jagadevan, S., Graham, N.J. & Thompson, I.P. (2013) Treatment of waste metalworking fluid by a hybrid ozone-biological process, *Journal of Hazardous Materials* 244-245, Elsevier, 394-402.

Jardim, W., Gimenez, S., Canela, M. & Moraes, S. (1993) Acute toxicity of Hg super (0) and Hg super (2+) ions to Escherichia coli, *Chemical Speciation and Bioavailability* 5, Science Reviews 2000, 97-100.

Kamble, S.P., Sawant, S.B. & Pangarkar, V.G. (2003) Batch and continuous photocatalytic degradation of benzenesulfonic acid using concentrated solar radiation, *Industrial & Engineering Chemistry Research* 42, American Chemical Society (ACS), 6705-6713.

Karunakaran, C., Dhanalakshmi, R., Gomathisankar, P. & Manikandan, G. (2010) Enhanced phenol-photodegradation by particulate semiconductor mixtures: Interparticle electron-jump, *Journal of Hazardous Materials* 176(1-3), Elsevier, 799-806.

Keen, O.S., Linden, K.G. (2013) Degradation of Antibiotic Activity during UV/H_2O_2 Advanced Oxidation and Photolysis in Wastewater Effluent, Environmental Science and Technology 47(22), *American Chemical Society* (ACS), 13020-13030.

Kelly, M.M., Arnold, W.A. (2012) Direct and Indirect Photolysis of the Phytoestrogens Genistein and Daidzein, Environmental Science & Technology 46(10), *American Chemical Society* (ACS), 5396-5403.

Khodja, A.A., Sehili, T., Pilichowski, J.F. & Boule, P. (2001) Photocatalytic degradation of 2-phenylphenol on TiO2 and ZnO in aqueous suspensions, *Journal of Photochemistry and Photobiology A: Chemistry* 141, Elsevier, 231-239.

Kim, K.H., Ihm, S.K. (2011) Heterogeneous catalytic wet air oxidation of refractory organic pollutants in industrial wastewaters: a review, *Journal of Hazardous Materials* 186, Elsevier, 16-34.

Kiwi, J., Pulgarin, C., Peringer, P. & Graetzel, M. (1993) Beneficial effects of homogeneous photo-Fenton pretreatment upon the biodegradation of anthraquinone sulfonate in waste water treatment, *Applied Catalysis B: Environmental* 3, Elsevier, 85-99.

Ko, C.H., Hsieh, P.H., Chang, M.W., Chern, J.M., Chiang, S.M. & Tzeng, C.J. (2009) Kinetics of pulp mill effluent treatment by ozone-based processes, *Journal of Hazardous Materials* 168(2-3), Elsevier, 875-881.

Kositzi, M., Poulios, I., Malato, S., Caceres, J. & Campos, A. (2004) Solar photocatalytic treatment of synthetic municipal wastewater, *Water Research 38*, the International Water Association (IWA), Elsevier, 1147-1154.

Ku, Y., Wang, W. & Shen, Y.S. (2000) Reaction behaviors of decomposition of monocrotophos in aqueous solution by UV and UV/O_3 processes, *Journal of Hazardous Materials* 72(1), Elsevier, 25-37.

Kuo, P., Chian, E. & Chang, B. (1978) Identification of end products resulting from ozonation of compounds commonly found in water, Ozone/Chlorine Dioxide Oxidation Products of Organic Materials, *International Ozone Institute, Ozone Press International,* 153-166.

Kusic, H., Peternel, I., Ukic, S., Koprivanac, N., Bolanca, T., Papic, S. & Bozic, A.L. (2011) Modeling of iron activated persulfate oxidation treating reactive azo dye in water matrix, *Chemical Engineering Journal* 172(1), Elsevier, 109-121.

Kusvuran, E., Gulnaz, O., Samil, A. & Erbil, M. (2010) Detection of double bond-ozone stoichiometry by an iodimetric method during ozonation processes, *Journal of Hazardous Materials* 175, Elsevier, 410-416.

Larue, C., Castillo-Michel, H., Sobanska, S., Trcera, N., Sorieul, S., Cecillon, L., Ouerdane, L., Legros, S. & Sarret, G. (2014) Fate of pristine TiO_2 nanoparticles and aged paint-containing TiO_2 nanoparticles in lettuce crop after foliar exposure, *Journal of Hazardous Materials* 273, Elsevier, 17-26.

Lee, Y., Gerrity, D., Lee, M., Bogeat, A.E., Salhi, E., Gamage, S., Trenholm, R.A., Wert, E.C., Snyder, S.A. & von Gunten, U. (2013) Prediction of Micropollutant Elimination during Ozonation of Municipal Wastewater Effluents: Use of Kinetic and Water Specific Information, Environmental Science & Technology 47(11), *American Chemical Society* (ACS), 5872-5881.

Lee, H.G., Sai-Anand, G., Komathi, S., Gopalan, A.L., Kang, S.W. & Lee, K.P. (2015) Efficient visible-light-driven photocatalytic degradation of nitrophenol by using graphene-encapsulated TiO_2 nanowires, *Journal of Hazardous Materials* 283, Elsevier, 400-409.

Lee, M., See, G. & Wynveen, R. (1977) Reaction Kinetics of UV/Ozone with Organic Compounds in Hospital Wastewater, *Symposium on Advanced Ozone Technology,* Toronto, Canada.

Legrini, O., Oliveros, E. & Braun, A. (1993) Photochemical processes for water treatment, *Chemical Reviews* 93, 671-698.

Lezcano, I., Nunez, N., Espino, M. & Gomez, M. (2000) Antibacterial Activity of Ozonized Sunflower Oil, Oleozón, Against Staphylococcus aureus and Staphylococcus epidermidis, *Ozone: Science & Engineering* 22, the International Ozone Association, Taylor & Francis, 207-214.

Lezcano, I., Perez rey, R., Gutierrez, M.S., Baluja, C. & Sanchez, E. (2001) Ozone inactivation of microorganisms in water: Gram positive Bacteria and yeast, *Ozone: Science & Engineering* 23, the International Ozone Association, Taylor & Francis,183-187.

Li, D., Huang, C.H. & Wang, Y.L. (2014) Effects of Combined UV and Chlorine Treatment on the Formation of Trichloronitromethane from Amine Precursors, *Environmental Science & Technology* 48(5), American Chemical Society (ACS), 2697-2705.

Li, K., Stefan, M.I. & Crittenden, J.C. (2007) Trichloroethene Degradation by UV/H_2O_2 Advanced Oxidation Process: Product Study and Kinetic Modeling, *Environmental Science & Technology* 41(5), American Chemical Society (ACS), 1696-1703.

Li, S.X., Wei, D., Mak, N.K., Cai, Z.W., Xu, X.R., Li, H.B. & Jiang, Y. (2009) Degradation of diphenylamine by persulfate: Performance optimization, kinetics and mechanism, *Journal of Hazardous Materials* 164(1), Elsevier, 26-31.

Li, Y., Wen, B., Ma, W., Chen, C. & Zhao, J. (2012) Photocatalytic Degradation of Aromatic Pollutants : A Pivotal Role of Conduction Band Electron in Distribution of Hydroxylated Intermediates, *Environmental Science & Technology 46*(9), American Chemical Society (ACS), 5093-5099.

Li, K., Zhang, H., Tang, T., Xu, Y., Ying, D., Wang, Y. & Jia, J. (2014) Optimization and application of TiO_2/Ti-Pt photo fuel cell (PFC) to effectively generate electricity and degrade organic pollutants simultaneously, *Water Research* 62, the International Water Association (IWA), Elsevier, 1-10.

Li, H., Zhou, S., Sun, Y. & Lv, J. (2010) Application of response surface methodology to the advanced treatment of biologically stabilized landfill leachate using Fenton's reagent, *Waste Management* 30(11), Elsevier, 2122-2129.

Liltved, H., Hektoen, H. & Efraimsen, H. (1995) Inactivation of bacterial and viral fish pathogens by ozonation or UV irradiation in water of different salinity, *Aquacultural Engineering* 14, Elsevier, 107-122.

Liang, C., Huang, C.F. & Chen, Y.J. (2008) Potential for activated persulfate degradation of BTEX contamination, *Water Research* 42(15), the International Water Association (IWA), Elsevier, 4091-4100.

Liang, C., Lee, I.L., Hsu, I.L., Liang, C.P. & Lin, Y.L. (2008) Persulfate oxidation of trichloroethylene with and without iron activation in porous media, Chemosphere 70(3), Elsevier, 426-435.

Liang, C., Liang, C.P. & Chen, C.C. (2009) pH dependence of persulfate activation by EDTA/Fe(III) for degradation of trichloroethylene, *Journal of Contaminant Hydrology* 106(3-4), Elsevier, 173-182.

Liang, C., Wang, Z.S. & Bruell, C.J. (2007) Influence of pH on persulfate oxidation of TCE at ambient temperature, *Chemosphere* 66(1), Elsevier, 106-113.

Liang, C., Wang, Z.S. & Mohanty, N. (2006) Influence of carbonate and chloride ions on persulfate oxidation of trichloroethylene at 20°C, *The Total Environment* 370(2-3), Elsevier, 271-277.

Lin, Y.T., Liang, C. & Chen, J.H. (2011) Feasibility study of ultraviolet activated persulfate oxidation of phenol, *Chemosphere* 82(8), Elsevier, 1168-1172.

Lin, J. & Yu, J.C. (1998) An investigation on photocatalytic activities of mixed TiO< sub> 2</sub>-rare earth oxides for the oxidation of acetone in air, *Journal of photochemistry and photobiology A: Chemistry* 116, Elsevier, 63-67.

Linsebigler, A.L., Lu, G. & Yates Jr, J.T. (1995) Photocatalysis on TiO_2 surfaces: principles, mechanisms, and selected results, *Chemical reviews* 95, 735-758.

Liu, X., Chen, G., Erwin, J.G., Adam, N.K. & Su, C. (2013) Release of phosphorous impurity from TiO_2 anatase and rutile nanoparticles in aquatic environments and its implications, *Water Research* 47(16), the International Water association (IWA), Elsevier, 6149-6156.

Lu, Y., Yu, H., Chen, S., Quan, X. & Zhao, H. (2012) Integrating Plasmonic Nanoparticles with TiO_2 Photonic Crystal for Enhancement of Visible-Light-Driven Photocatalysis,

Environmental Science & Technology 46(3), American Chemical Society (ACS), 1724-1730.

Mahamuni, N.N., Adewuyi, Y.G. (2010) Advanced oxidation processes (AOPs) involving ultrasound for waste water treatment: a review with emphasis on cost estimation, *Ultrasonics Sonochemistry* 17, Elsevier, 990-1003.

Makino, K., Mossoba, M.M. & Riesz, P. (1982) Chemical effects of ultrasound on aqueous solutions. Evidence for hydroxyl and hydrogen free radicals (cntdot. OH and cntdot. H) by spin trapping, *Journal of the American Chemical Society* 104, 3537-3539.

Maness, P.C., Smolinski, S., Blake, D.M., Huang, Z., Wolfrum, E.J. & Jacoby, W.A. (1999) Bactericidal activity of photocatalytic TiO_2 reaction: toward an understanding of its killing mechanism, *Applied and environmental microbiology* 65, American Society for Microbiology, 4094-4098.

Marc, G., Augugliaro, V., Lopez-Muoz, M.J., Martin, C., Palmisano, L., Rives, V., Schiavello, M., Tilley, R.J. & Venezia, A.M. (2001) Preparation characterization and photocatalytic activity of polycrystalline ZnO/TiO_2 systems. 1. surface and bulk characterization, *The Journal of Physical Chemistry* B 105, American Chemical Society (ACS), 1026.

Marci, G., Augugliaro, V., Lopez-Munoz, M.J., Martin, C., Palmisano, L., Rives, V., Schiavello, M., Tilley, R.J. & Venezia, A.M. (2001) Preparation characterization and photocatalytic activity of polycrystalline ZnO/TiO_2 systems. 2. Surface, bulk characterization, and 4-nitrophenol photodegradation in liquid-solid regime, *The Journal of Physical Chemistry* B 105, American Chemical Society (ACS), 1033-1040.

Mare, M., Waldner, G., Bauer, R., Jacobs, H. & Broekaert, J. (1999) Degradation of nitrogen containing organic compounds by combined photocatalysis and ozonation, *Chemosphere* 38, Elsevier, 2013-2027.

Margot, J., Kienle, C., Magnet, A., Weil, M., Rossi, L., de Alencastro, L.F., Abegglen, C., Thonney, D., Chevre, N., Shaerer, M. & Barry, D.A. (2013) Treatment of micropollutants in municipal wastewater: Ozone or powdered activated carbon?, *Science of The Total Environment* 461-462, Elsevier, 480-498.

Martin, M.M.B., Sanchez Perez, J.A., Casas Lopez, J.L., Oller, I. & Malato Rodriguez, S. (2009) Degradation of a four-pesticide mixture by combined photo-Fenton and biological oxidation, *Water Research* 43(3), the International Water Association (IWA), Elsevier, 653-660.

Mendez-Hermida, F., Ares-Mazas, E., McGuigan, K.G., Boyle, M., Sichel, C. & Fernandez-Ibanez, P. (2007) Disinfection of drinking water contaminated with< i> Cryptosporidium parvum</i> oocysts under natural sunlight and using the photocatalyst TiO_2, *Journal of Photochemistry and Photobiology B: Biology* 88, Elsevier, 105-111.

Merenyi, G., Lind, J., Naumov, S. & von Sonntag, C. (2010) Reaction of Ozone with Hydrogen Peroxide (Peroxone Process): A Revision of Current Mechanistic Concepts Based on Thermokinetic and Quantum-Chemical Consideration, *Environmental Science & Technology* 44(9), American Chemical Society (ACS), 3505-3507.

Michael, I., Hapeshi, E., Michael, C., Varela, A.R., Kyriakou, S., Manaia, C.M. & Fatta-Kassinos, D. (2012) Solar photo-Fenton process on the abatement of antibiotics at a pilot scale: Degradation kinetics, ecotoxicity and phytotoxicity assessment and removal of antibiotic resistant enterococci, *Water Research* 46(17), the International Water Association (IWA), Elsevier, 5621-5634.

Mills, A., Belghazi, A. & Rodman, D. (1996) Bromate removal from drinking water by semiconductor photocatalysis, *Water Research* 30(9), the International Water Association (IWA), Elsevier, 1973-1978.

Mills, A., Davies, R.H. & Worsley, D. (1993) Water purification by semiconductor photocatalysis, *Chemical Society Reviews* 22, Royal Institute of Chemistry, 417-425.

Miralles-Cuevas, S., Oller, I., Sanchez-Perez, J.A. & Malato, S. (2014) Removal of pharmaceuticals from MWTP effluent by nanofiltration and solar photo-Fenton using two different iron complexes at neutral pH, *Water Research* 64, the International Water Association (IWA), Elsevier, 23-31.

Moreau, M., Orange, N. & Feuilloley, M. (2008) Non-thermal plasma technologies: new tools for bio-decontamination, *Biotechnology Advances* 26, Elsevier, 610-617.

Munoz, R., Guieysse, B. (2006) Algal–bacterial processes for the treatment of hazardous contaminants: a review, *Water Research* 40, the International Water Association, Elsevier, 2799-2815.

Mysore, C., Leparc, J., Lake, R., Agutter, P. & Prevost, M. (2004) Comparing static mixer performances at pilot and full scale for ozonation, inactivation of Bacillus subtilis, and bromate formation in water treatment, *Ozone: Science & Engineering 26*, the International Ozone Association, Taylor & Francis, 207-215.

Navarro, P., Sarasa, J., Sierra, D., Esteban, S. & Ovelleiro, J. (2005) Degradation of wine industry wastewaters by photocatalytic advanced oxidation, *Water Science & Technology* 51, IWA Publishing, 113-120.

Nfodzo, P., Choi, H. (2011) Triclosan decomposition by sulfate radicals: Effects of oxidant and metal doses, *Chemical Engineering Journal* 174(2-3), Elsevier, 629-634.

Noethe, T., Fahlenkamp, H. & von Sonntag, C. (2009) Ozonation of wastewater: Rate of ozone consumption and hydroxyl radical yield, *Environmental Science & Technology* 43(15), American Chemical Society (ACS), 5990-5995.

Ollis, D. (1986) *Heterogeneous photocatalysis for water purification: Prospects and problems. Homogeneous and heterogeneous photocatalysis,* Springer.

Ollis, D.F., Al-Ekabi, H. (1993) *Photocatalytic purification and treatment of water and air: proceedings of the 1st International Conference on* TiO_2 *Photocatalytic Purification and Treatment of Water and Air,* London, Ontario, Canada, 8-13 November, 1992, Elsevier.

Ollis, D.F. (1985) Contaminant degradation in water, *Environmental Science & Technology* 19, American Chemical Society (ACS), 480-484.

Ollis, D.F., Pelizzetti, E. & Serpone, N. (1991) Photocatalyzed destruction of water contaminants, *Environmental Science & Technology* 25, American Chemical Society (ACS), 1522-1529.

Oya, M., Kosaka, K., Asami, M. & Kunikane, S. (2008) Formation of< i> N</i>-nitrosodimethylamine (NDMA) by ozonation of dyes and related compounds, *Chemosphere* 73, Elsevier, 1724-1730.

Pareek, V.K., Adesina, A.A. (2004) Light intensity distribution in a photocatalytic reactor using finite volume, AIChE journal 50, *American Institute of Chemical Engineers,* Wiley, 1273-1288.

Park, H.S., Hwang, T.M., Kang, J.W., Choi, H. & Oh, H.J. (2001) Characterization of raw water for the ozone application measuring ozone consumption rate, *Water Research* 35, the International Water Association (IWA), Elsevier, 2607-2614.

Parsons, S. (2004) Advanced oxidation processes for water and wastewater treatment, IWA publishing.

Perez, J.A.S., Sanchez, I.M.R., Carra, I., Reina, A.C., Lopez, J.L.C. & Malato, S. (2013) Economic evaluation of a combined photo-Fenton/MBR process using pesticides as model pollutant. Factors affecting costs, *Journal of Hazardous Materials* 244-245, Elsevier, 195-203.

Peternel, I., Koprivanac, N. & Kusic, H. (2006) UV-based processes for reactive azo dye mineralization, *Water Research 40, the International Water Association* (IWA), Elsevier, 525-532.

Peyton, G.R., Glaze, W.H. (1987) Mechanism of photolytic ozonation, ACS Symposium Series.

Peyton, G.R., Huang, F.Y., Burleson, J.L. & Glaze, W.H. (1982) Destruction of pollutants in water with ozone in combination with ultraviolet radiation. 1. General principles and oxidation of tetrachloroethylene, *Environmental Science & Technology 16, American Chemical Society* (ACS), 448-453.

Pichat, P. (2003) Photocatalytic degradation of pollutants in water and air: basic concepts and applications, *Environmental Science and Pollution Control Series,* 77-120.

Pignatello, J.J. (1992) Dark and photoassisted iron (3+)-catalyzed degradation of chlorophenoxy herbicides by hydrogen peroxide, *Environmental Science & Technology 26, American Chemical Society* (ACS), 944-951.

Pignatello, J.J., Oliveros, E. & MacKay, A. (2006) Advanced oxidation processes for organic contaminant destruction based on the Fenton reaction and related chemistry, *Critical reviews in Environmental Science & Technology* 36, 1-84.

Pocostales, J.P., Sein, M.M., Knolle, W., von Sonntag, C. & Schmidt, T.C. (2010) Degradation of Ozone-Refractory Organic Phosphates in Wastewater by Ozone and Ozone/Hydrogen Peroxide (Peroxone): The Role of Ozone Composition by Dissolved Organic Matter, *Environmental Science & Technology* 44(21), American Chemical Society (ACS), 8248-8253.

Popiel, S., Witkiewicz, Z. & Chrzanowski, M. (2008) Sulfur mustard destruction using ozone, UV, hydrogen peroxide and their combination, *Journal of Hazardous Materials* 153(1-2), Elsevier, 37-43.

Popov, E., Mametkuliyev, M., Santoro, D., Liberti, L. & Eloranta, J. (2010) Kinetics of UV/H_2O_2 Advanced Oxidation in the Presence of Alcohols: The Role of Carbon Centered Radicals, *Environmental Science & Technology* 44(20), American Chemical Society, 7827-7832.

Pirkanniemi, K., Sillanpaeae, M. (2002) Heterogeneous water phase catalysis as an environmental application: a review, *Chemosphere* 48, Elsevier, 1047-1060.

Prengle Jr, H. (1977) *Evolution of the Ozone/UV Process for Wastewater Treatment,* IOA/EPA Colloquium on Wastewater Treatment & Disinfection with Ozone, US Environmental Protection Agency, Municipal Environmental Research Laboratory, Cincinnati, OH.

Prengle J., H., Mank, C. & Payne, J. (1976) Ozone-UV oxidation of pesticides in aqueous solution, *Technical paper presented at the IOI Meeting,* Cincinnati, OH, USA.

Prengle, J., Hewes, C. & Mank, C. (1975) Oxidation of refractory materials by ozone with ultraviolet radiation, *Proceeding of the 2nd International Symposium,* Ozone Institute, Incorporated, Cleveland, OH, USA.

Pulgarin, C., Kiwi, J. (1996) Overview on photocatalytic and electrocatalytic pretreatment of industrial non-biodegradable pollutants and pesticides, *CHIMIA International Journal for Chemistry 50, Swiss Chemical Society* (SCS), 50-55.

Ranjit, K. & Viswanathan, B. (1997) Synthesis, characterization and photocatalytic properties of iron-doped TiO_2 catalysts, *Journal of Photochemistry and Photobiology A: Chemistry* 108, Elsevier, 79-84.

Rao, Y.F., Chu, W. (2010) Degradation of linuron by UV, ozonation, and UV/O_3 processes – Effect of anions and reaction mechanism, *Journal of Hazardous Materials* 180(1-3), Elsevier, 514-523.

Rice, R.G. (1996) Applications of ozone for industrial wastewater treatment-a review, *Ozone: Science & Engineering 18, the International Ozone Association,* Taylor & Francis, 477-515.

Robertson, P.K.J., Robertson, J.M.C & Bahnemann, D.W. (2012) Removal of microorganisms and their chemical metabolites from water using semiconductor photocatalysis, *Journal of Hazardous Materials* 211-212, Elsevier, 161-171.

Rodriguez-Romoand, L.A. & Yousef, A.E. (2005) Inactivation of Salmonella enterica serovar Enteritidis on shell eggs by ozone and UV radiation, *Journal of Food Protection® 68, JFP online, International Association for Food Production®*, 711-717.

Rosal, R., Rodriguez, A., Perdigon-Melon, J.A., Petre, A. & Garcia-Calvo, E. (2009) Oxidation of dissolved organic matter in the effluent of a sewage treatment plant using ozone combined with hydrogen peroxide (O_3/H_2O_2), *Chemical Engineering Journal* 149(1-3), Elsevier, 311-318.

Rosal, R., Rodriguez, A., Perdigon-Melon, J.A., Mezcua, M., Hernando, M.D., Leton, P., Garcia-Calvo, E., Aguera, A. & Fernandez-Alba, A.R. (2008) Removal of pharmaceuticals and kinetics of mineralization by O_3/H_2O_2 in a biotreated municipal wastewater, *Water Research* 42(14), the International Water Association (IWA), Elsevier, 3719-3728.

Rosenfeldt, E.J., Linden, K.G. (2007) The $R_{OH,UV}$ Concept to Characterize and The Model UV/H_2O_2 Process in Natural Waters, Environmental Science & Technology 41(7), *American Chemical Society,* 2548-2553.

Rumi, I.R., Mitsugu, I.H. & Takashima, K. (2005) Photocatalytic degradation of imazethapyr herbicide at TiO_2/H_2O interface, *Chemosphere* 58, Elsevier, 1461-1469.

Safa, S., Soucy, G. (2014) Liquid and solution treatment by thermal plasma: a review, *International Journal of Environmental Science and Technology* (IJEST) 11, Springer, 1165-1188.

Safarzadeh-Amiri, A., Bolton, J.R. & Cater, S.R. (1996) Ferrioxalate-mediated solar degradation of organic contaminants in water, *Solar Energy* 56, Elsevier, 439-443.

Safarzadeh-Amiri, A., Bolton, J.R. & Cater, S.R. (1997) Ferrioxalate-mediated photodegradation of organic pollutants in contaminated water, *Water Research 31, the International Water Association* (IWA), Elsevier, 787-798.

Santoro, D., Raisee, M., Moghaddami, M., Ducoste, J., Sasges, M., Liberti, L. & Notarnicola, M. (2010) Modeling Hydroxyl Radical Distribution and Trialkyl Phosphates Oxidation in UV-H_2O_2 Photoreactors Using Computational Fluid Dynamics, Environmental Science & Technology 44(16), *American Chemical Society* (ACS), 6233-6241.

Sarathy, S.R., Mohseni, M. (2007) The Impact of UV/H_2O_2 Sdvanced Oxidation on Molecular Size Distribution of Chromophoric Natural Organic Matter, *Environmental Science & Technology* 41(24), American Chemical Society (ACS), 8315-8320.

Saritha, P., Aparna, C., Himabindu, V. & Anjaneyulu, Y. (2007) Comparison of various advanced oxidation processes for the degradation of 4-chloro-2 mitrophenol, *Journal of Hazardous Materials* 149(3), Elsevier, 609-614.

Schmidt, C.K., Brauch, H.J. (2008) N,N-dimethylsulfamide as precursor for N-nitrosodimethylamine (NDMA) formation upon ozonation and its fate during drinking water treatment, Environmental Science & Technology 42, *American Chemical Society* (ACS), 6340-6346.

Serpone, N., Maruthamuthu, P., Pichat, P., Pelizzetti, E. & Hidaka, H. (1995) Exploiting the interparticle electron transfer process in the photocatalysed oxidation of phenol, 2-chlorophenol and pentachlorophenol: chemical evidence for electron and hole transfer between coupled semiconductors, *Journal of Photochemistry and Photobiology A: Chemistry* 85, Elsevier, 247-255.

Shang, N.C., Chen, Y.H., Ma, H.W., Lee, C.W., Chang, C.H., Yu, Y.H. & Lee, C.H. (2007) Oxidation of methyl methacrylate from semiconductor wastewater by O_3 and O_3/UV processes, *Journal of Hazardous Materials* 147(1-2), Elsevier, 307-312.

Sharpless, C.M., Linden, K.G. (2003) Experimental and Model Comparisons of Low- and Medium-Pressure Hg Lamps for the Direct and H_2O_2 Assisted UV Photodegradation of N-Nitrosodimethylamine in Simulated Drinking Water, Environmental Science & Technology 37(9), *American Chemical Society* (ACS), 1933-1940.

Shiraishi, F., Nakasako, T. & Hua, Z. (2003) Formation of hydrogen peroxide in photocatalytic reactions, The Journal of Physical Chemistry A 107, *American Chemical Society* (ACS), 11072-11081.

Shu, H.Y. & Chang, M.C. (2005a) Decolorization effects of six azo dyes by O_3, UV/O_3 and UV/H_2O_2 processes, *Dyes and Pigments* 65, Elsevier, 25-31.

Shu, H.Y., Chang, M.C. (2005b) Pre-ozonation coupled with UV/H_2O_2 process for the decolorization and mineralization of cotton dyeing effluent and synthesized CI Direct Black 22 wastewater, *Journal of Hazardous Materials* 121, Elsevier, 127-133.

Shu, H.Y. & Huang, C.R. (1995) Degradation of commercial azo dyes in water using ozonation and UV enhanced ozonation process, *Chemosphere* 31, Elsevier, 3813-3825.

Silva, T.F.C.V., Fonseca, A., Saraiva, I., Vilar, V.J.P. & Boaventura, R.A.R. (2013a) Biodegradability enhancement of a leachate after biological lagooning using a solar driven photo-Fenton reaction, and further combination with an activated sludge biological process at pre-industrial scale, *Water Research 47(10), the International Water Association* (IWA), Elsevier, 3543-3557.

Silva, T.F.C.V., Silva, M.E.F., Cunha-Queda, A.C., Fonseca, A., Saraiva, I., Sousa, M.A., Goncalves, C., Alpendurada, M.F., Boaventura, R.A.R. & Vilar, V.J.P. (2013b) Multistage treatment system for raw leachate from sanitary landfill combining biological nitrification-denitrification/solar photo-Fenton/biological processes, at a scale close to industrial-Biodegradability enhancement and evolution profile of trace pollutants, *Water Research* 47(16), the International Water association (IWA), Elsevier, 6167-6186.

Somensi, C.A., Simionatto, E.L., Bertoli, S.L., Wisniewski Jr, A. & Radetski, C.M. (2010) Use of ozone in a pilot-scale plant for textile wastewater pre-treatment: Physico-chemical efficiency, degradation by-products identification and environmental toxicity of treated wastewater, *Journal of Hazardous Materials* 175(1-3), Elsevier, 235-240.

Spanggord, R.J., Yao, D. & Mill, T. (2000) Kinetics of Aminodinitrotoluene Oxidations with Ozone and Hydroxyl Radical, *Environmental Science & Technology* 34(3), American Chemical Society (ACS), 450-454.

Staehelin, J. & Hoigne, J. (1982) Decomposition of ozone in water: rate of initiation by hydroxide ions and hydrogen peroxide, *Environmental Science & Technology* 16, American Chemical Society (ACS), 676-681.

Stafford, U., Gray, K.A. & Kamat, P.V. (1996) Photocatalytic degradation of organic contaminants: Halophenols and related model compounds, *Heterogeneous Chemistry Reviews* 3, Wiley, 77-104.

Stefan, M.I. & Bolton, J.R. (1999) Reinvestigation of the Acetone Degradation Mechanism in Dilute Aqueous Solution by the UV/H_2O_2 Process, *Environmental Science & Technology* 33, American Chemical Society (ACS), 870-873.

Suarez, S., Dodd, M.C., Omil, F. & Von Gunten, U. (2007) Kinetics of triclosan oxidation by aqueous ozone and consequent loss of antibacterial activity: relevance to municipal wastewater ozonation, *Water Research* 41, the International Water Association (IWA), Elsevier, 2481-2490.

Taube, H. (1957) Photochemical reactions of ozone in solution, *Transactions of the Faraday Society* 53, 656-665.

Tomiyasu, H., Fukutomi, H. & Gordon, G. (1985) Kinetics and mechanism of ozone decomposition in basic aqueous solution, *Inorganic Chemistry* 24, American Chemical Society (ACS), 2962-2966.

Tong, T., Binh, C.T.T., Kelly, J.J., Gaillard, J.F. & Gray, K.A. (2013) Cytotoxicity of commercial nano-TiO_2 to Escherichia coli assessed by high-throughout screening: Effects of environmental factors, *Water Research* 47(7), the International Water Association (IWA), Elsevier, 2352-2362.

Trovo, A.G., Nogueira, R.F.P., Aguera, A., Fernandez-Alba, A.R., Sirtori, C. & Malato, S. (2009) Degradation of sulfamethoxazole in water by solar photo-Fenton. Chemical and toxicological evaluation, *Water Research* 43(16), the International Water Association (IWA), Elsevier, 3922-3931.

Umar, M., Roddick, F., Fan, L. & Abdul Aziz, H. (2013) Application of ozone for the removal of bisphenol A from water and wastewater – A review, *Chemosphere* 90(8), Elsevier, 2197-2207.

Ventura, A., Jacquet, G., Bermond, A. & Camel, V. (2002) Electrochemical generation of the Fenton's reagent: application to atrazine degradation, Water Research 36, *the International Water Association* (IWA), Elsevier, 3517-3522.

Villasenor, J., Mansilla, H.D. (1996) Effect of temperature on kraft black liquor degradation by ZnO-photoassisted catalysis, *Journal of Photochemistry and Photobiology A: Chemistry* 93, Elsevier, 205-209.

Vinodgopal, K., Wynkoop, D.E. & Kamat, P.V. (1996) Environmental Photochemistry on Semiconductor Surfaces: Photosensitized Degradation of a Textile Azo Dye, Acid Orange 7, on TiO_2 Particles Using Visible Light, *Environmental Science & Technology* 30(5), American Chemical Society (ACS), 1660-1666.

Vinu, R., Madras, G. (2008) Kinetics of Simultaneous Photocatalytic Degradation of Phenolic Compounds and Reduction of Metal Ions with Nano-TiO_2, *Environmental Science & Technology* 42(3), American Chemical Society (ACS), 913-919.

Vinu, A., Sawant, D.P., Ariga, K., Hartmann, M. & Halligudi, S. (2005) Benzylation of benzene and other aromatics by benzyl chloride over mesoporous AlSBA-15 catalysts, *Microporous and mesoporous materials* 80, Elsevier, 195-203.

Wang, X., Chen, S., Gu, X. & Wang, K. (2009) Pilot study on the advanced treatment of landfill leachate using a combined coagulation, fenton oxidation and biological aerated filter process, *Waste Management* 29(4), Elsevier, 1354-1358.

Wang, K., Guo, J., Yang, M., Junji, H. & Deng, R. (2009) Decomposition of two haloacetic acids in water using UV radiation, ozone and advanced oxidation processes, *Journal of Hazardous Materials* 162(2-3), Elsevier, 1243-1248.

Wang, X., Han, J., Chen, Z., Jian, L., Gu, X. & Lin, C.J. (2012) Combined processes of two-stage Fenton-biological anaerobic filter-biological aerated filter for advanced treatment of landfill leachate, *Waste Management* 32(12), Elsevier, 2401-2405.

Wang, X.Q., Liu, C.P., Yuan, Y. & Li, F.B. (2014) Arsenite oxidation and removal driven by a bio-electro-Fenton process under neutral pH conditions, *Journal of Hazardous Materials* 275, Elsevier, 200-209.

Wang, J., Ried, A., Wieland, A., Zhang, Y., Chen, M., Xie, R., Ang, W. S., Duarah, A., Pang, T. & Zhang, L. (2012) Ozone/H_2O_2 AOP Pilot Testing at a Waterworks with Dynamic Raw Water Characteristics in Singapore, *Water Practice & Technology* 7, IWA Publishing.

Wang, X.J., Song, Y. & Mai, J.S. (2008) Combined Fenton oxidation and aerobic biological processes for treating a surfactant wastewater containing abundant sulfate, *Journal of Hazardous Materials* 160(2-3), Elsevier, 344-348.

Weiss, J. (1944) Radiochemistry of aqueous solutions, *Nature* 153, 48-50.

Wong, C., Chu, W. (2003) The hydrogen peroxide-assisted photocatalytic degradation of alachlor in TiO_2 suspensions, Environmental Science & Technology 37, *American Chemical Society* (ACS), 2310-2316.

Wu, M., Jin, Y., Zhao, G., Li, M. & Li, D. (2010) Electrosorption-promoted Photodegradation of Opaque Wastewater on a Novel TiO_2/Carbon Aerogel Electrode, *Environmental Science & Technology* 44(5), American Chemical Society (ACS), 1780-1785.

Xie, B., Lv, Z., Lv, B.Y. & Gu, Y.X. (2010) Treatment of mature landfill leachate by biofilters and Fenton oxidation, *Waste Management* 30(11), Elsevier, 2108-2112.

Xu, B., Chen, Z., Qi, F., Ma, J. & Wu, F. (2010) Comparison of N-nitrosodiethylamine degradation in water by UV irradiation and UV/O_3: Efficiency, product and mechanism, *Journal of Hazardous Materials* 279, Elsevier, 976-982.

Xu, X.R., Li, X.Z. (2010) Degradation of azo dye orange G in aqueous solution by persulfate with ferrous ion, *Separation and Purification Technology* 72(1), Elsevier, 105-111.

Xu, S.C., Pan, S.S., Xu, Y., Luo, Y.Y., Zhang, Y.X. & Li, G.H. (2015) Efficient removal of Cr(VI) from wastewater under sunlight by Fe(II)-doped TiO_2 spherical shell, *Journal of Hazardous Materials* 283, Elsevier, 7-13.

Yamazaki, S., Matsunaga, S. & Hori, K. (2001) Photocatalytic degradation of trichloroethylene in water using TiO_2 pellets, *Water Research* 35, the International Water Association (IWA), Elsevier, 1022-1028.

Yan, T., Long, J., Shi, X., Wang, D., Li, Z. & Wang, X. (2010) Efficient Photocatalytic Degradation of Volatile Organic Compounds by Porous Indium Hydroxide Nanocrystals, *Environmental Science & Technology* 44(4), American Chemical Society (ACS), 1380-1385.

Yang, S., Wang, P., Yang, X., Wei, G., Zhang, W. & Shan, L. (2009) A novel advanced oxidation process to degrade organic pollutants in wastewater: Microwave-activated persulfate oxidation, *Journal of Environmental Science* 21(9), Elsevier, 1175-1180.

Yang, S., Yang, X., Shao, X., Niu, R. & Wang, L. (2011) Activated carbon catalyzed persulfate oxidation of Azo dye acid orange 7 at ambient temperature, *Journal of Hazardous Materials* 186(1), Elsevier, 659-666.

Yao, H., Sun, P., Minakata, D., Crittenden, J.C. & Huang, C.H. (2013) Kinetics and Modeling of Degradation of Ionophore Antibiotics by UV and UV/H_2O_2, *Environmental Science & Technology* 47(9), American Chemical Society (ACS), 4581-4589.

Yao, K., Wang, D., Chang, C., Weng, K., Yang, L., Lee, S., Cheng, T. & Hwang, C. (2007) Photocatalytic disinfection of phytopathogenic bacteria by dye-sensitized TiO_2 thin film activated by visible light, *Surface and Coatings Technology* 202, Elsevier, 1329-1332.

Yeber, M., Rodriguez, J., Freer, J., Baeza, J., Duran, N. & Mansilla, H. D. (1999) Advanced oxidation of a pulp mill bleaching wastewater, *Chemosphere* 39, Elsevier, 1679-1688.

Zhang, A.Y., Long, L.L., Liu, C., Li, W.W. & Yu, H.Q. (2014) Electrochemical degradation of refractory pollutants using TiO_2 single crystals exposed by high-energy {001} facets, *Water Research* 66, the International Water Association (IWA), Elsevier, 273-282.

Zhao, Y.Y., Boyd, J.M., Woodbeck, M., Andrews, R.C., Qin, F., Hrudey, S.E. & Li, X.F. (2008) Formation of N-nitrosamines from eleven disinfection treatments of seven different surface waters, *Environmental Science & Technology* 42, American Chemical Society (ACS), 4857-4862.

Zhang, Z., Yu, F., Huang, L., Jiatieli, J., Li, Y., Song, L., Yu, N. & Dionysion, D.D. (2014) Confirmation of hydroxyl radicals generated in the presence of TiO_2 supported on AC under microwave irradiation, *Journal of Hazardous Materials* 278, Elsevier, 152-157.

Zheng, T., Arnold, W.A. (2013) Pesticide Photolysis in Prairie Potholes: Probing Photosensitized Processes, *Environmental Science & Technology* 47(13), American Chemical Society (ACS), 6735-6745.

In: Enhanced Electrochemical Advanced Oxidation … ISBN: 978-1-63482-089-9
Editor: Gagik Badalians Gholikandi

Chapter 3

ELECTROCHEMICAL ADVANCED OXIDATION PROCESSES BASED ON FENTON'S REACTION

Gagik Badalians Gholikandi[1]* and Hamidreza Masihi[2]
[1]Faculty of Water Engineering and Environment,
Shahid Beheshti University, A.C., Tehran, Iran
[2]Faculty of Water Engineering and Environment,
Shahid Beheshti University, A.C., Tehran, Iran

INTRODUCTION

Over the past decades, advanced oxidation processes (AOPs) have attracted increasing interest as promising efficient methods for removing persistent organic contaminants from water and wastewater (Kang et al., 1987; Beltran et al., 1997; Andreozzi et al., 1999; Tarr 2003; Gogate and Pandit 2004; Parsons 2004; Brillas et al., 2006; Canizares et al., 2006; Garrido et al., 2007; Laine and Cheng 2007, Lei and Dai 2007; Brillas et al., 2009; Bach et al., 2009; Gunagi et al., 2010; Huang et al., 2010; Bacardit et al., 2010; Boudreau et al., 2010; De Luis et al., 2011; Ahmed et al., 2011; Rojas et al., 2011; Vijayalakshmi et al., 2011; Real et al., 2012; He et al., 2012;Woisetschlaeger et al., 2013; Yuan et al., 2013; Li et al., 2013; Krzywicka and Kwarciak 2014; Dong et al., 2014; Zhang et al., 2014; Abeish et al., 2014; Rodrigo et al., 2014; Dai et al., 2014; Choi and Chung 2014; Milovac et al., 2014; Lee et al., 2014; Rossi et al., 2014; Guo et al., 2014). Multidisciplinary researches have been carried out to investigate and evaluate various related topics such as process fundamentals, determination of system kinetics and mechanisms, exploration of novel materials, and process modelling, integration and scale up (Comninellis et al., 2008). In this regard, electrochemical advanced oxidation processes (EAOPs) have emerged as novel environmental-friendly and effective treatment technologies for the elimination of a several organic contaminants (Sires et al., 2014, Martinez-Huitle 2014). Considerable validation of these methods has been performed at both the bench-scale and pilot-scale. Although a promising new technology, the mechanisms

* Corresponding Author, Hakimieh- Tehranpars, Shahid Abbaspour Blvd., PO Code 16589-53571, PO Box 16765-1719, Tehran, Iran. Email: g.badalians@yahoo.com.

involved in the oxidation of organic compounds during electrochemical advanced oxidation processes (EAOP) and the corresponding environmental impacts have not been completely addressed until now. Important challenges related to electrochemical advanced oxidation processes are belonging to poisonous byproducts production and low electro-active areas (Chaplin 2014). Energy consumption and operational costs are also related key concerns. A combination of electrochemical advanced oxidation processes with biological treatment methods is an interesting solution in this regard. Consequently, the removal efficiency could be maximized, while the operational costs would be optimized (Ganzenko et al., 2014).

Electrochemical advanced oxidation processes have been employed effectively to remove refractory organics from landfill leachate and wastewaters containing organics, especially aromatic compounds (Nidheesh and Gandhimathi, 2012; Laiju et al., 2014). Nowadays, the application of this technology is one of the compromises for landfill leachate treatment because of its several various advantages (Muhammad et al., 2010).

The typical advanced oxidation processes can be divided into homogenous systems with irridation (O_3/UV, H_2O_2/UV, Ultrasounds (US), UV/US, electron beam, H_2O_2/US processes, H_2O_2/Fe^{2+}/UV (photo-Fenton)), homogenous systems without irradiation (O_3/H_2O_2, O_3/OH^-, H_2O_2/Fe^{2+} (Fenton's)), heterogeneous systems with irradiation (TiO_2/O_2/UV, TiO_2/H_2O_2/UV), and heterogeneous system without irradiation (electro-Fenton) (Huang et al., 1993).

In the recent years, a large variety of electrochemical advanced oxidation processes has been proposed. The most investigated method is the electro-Fenton process. Combined electro-Fenton methods have also been considered for wastewater remediation, including peroxi-coagulation, photoelectron-Fenton, solar photoelectron-Fenton, photoperoxi-coagulation, photoelectrochemical electro-Fenton, sonoelectro-Fenton, and cathodic generation of Fe^{2+} as well as electrochemical and combined Fenton processes with hydrogen peroxide added to the solution or produced indirectly, including Fered-Fenton, electrochemical peroxidation, anodic Fenton treatment, photocombined methods, bio-electrochemical peroxidation, and plasma-assisted treatments (Brillas et al., 2009; Sires et al., 2014).

The basic affecting parameters of the electrolysis application are as follows (Pletcher and Walsh 1982, Panizza 2010):

1. *Electrode potential and current density:* Reactions occurrence and their rates control, the determination of the process efficiency.
2. *Current distribution:* Determines the spatial distribution of the reactants consumption and hence, it must be as homogeneous as possible.
3. *Mass-transport regime:* High efficiency and greate uniformity of contaminants concentration in the reaction zone close to electrode surface results from the high mass-transport.
4. *Cell design:* Various related parameters such as cell dimension, divided or undivided cell and specification of the electrode employed affect the occurrence of the electrochemical process.
5. *Electrolysis medium:* The kind of electrolyte and its concentration, pH, and temperature are the related considerable parameters.
6. *Electrode materials:* The suitable electrode material for the decomposition of organic contaminants should have the necessary stability in the electrolysis medium and it

must be cost effective. The high activity according to organic oxidation and low activity related to secondary reactions is a significant characteristic in this regard.

The electrochemical mineralization of soluble organics in water should be occured at very low thermodynamic potentials as follows (the reaction 1) (Kapalka et al., 2010):

$$2H_2O \rightarrow O_2 + 4H^+ + 4e^- \quad (1)$$

It never exceeds 0.2 V/SHE. In contrast, the kinetics of the electrochemical mineralization is very slow.

Anodic oxidation reactions, i.e., electrochemical oxygen transfer reaction (EOTR), occur simultaneously with the transfer of oxygen from water to the reaction products. An example of this reaction is the anodic electrochemical incineration of phenol (the reaction 2) (Panizza et al., 2008):

$$C_6H_5OH + 11H_2O \rightarrow 6CO_2 + 28H^+ + 28e^- \quad (2)$$

In this reaction, water is the source of oxygen atoms for the total oxidation of phenol to CO_2 at the anode. In this regard, the released protons are discharged at the cathode to dihydrogen (the reaction 3) (Panizza et al., 2008):

$$2H^+ + 2e^- \rightarrow H_2 \quad (3)$$

Adsorbed hydroxyl radicals are produced as a result of the water discharge on the electrode based on activation mechanism of water by electrolytic discharge in acid media under standard conditions; thermodynamic potential equals 1.23 V/SHE (the reaction 4). The essential intermediate for the evolution of oxygen is the hydroxyl radical (the reaction 5) (Panizza et al., 2008, Kapalka et al., 2010; Panizza 2010).

$$H_2O + M \rightarrow M\,(\bullet OH) + H^+ + e^- \quad (4)$$

$$M\,(\bullet OH) \rightarrow M + \tfrac{1}{2}\,O_2 + H^+ + e^- \quad (5)$$

The electrochemical oxygen transfer reaction between an organic compound R and the hydroxyl radicals occurs close to the anode's surface (the reaction 6) (Kapalka et al., 2010):

$$R + M\,(\bullet OH)_{n/2} \rightarrow M + \text{oxidation products} + n/_2\,H^+ + n/_2\,e^- \quad (6)$$

n = the electrons number involved in oxidation reaction of the organic compound.

The activity of the hydroxyl radicals is closely related to their interaction with the electrode surface M.

The related electrochemical parameters are as follows:

1. Instantaneous Current Efficiency (ICE)

Determines the current efficiency of the organics mineralization. In this regard, The chemical oxygen demand (COD) of the electrolyte has been considered for the ICE's determination employing the following equeation (equation 7) (Kapalka et al., 2010):

$$ICE_{COD} = FV/8I = [(COD)_t - (COD)_{t+\Delta t}] / \Delta t \qquad (7)$$

$(COD)_t$ and $(COD)_{t+\Delta t}$ = the chemical oxygen demands (mol O_2 m^{-3}) at times t and t+Δt (s), I = the applied current (A),

F = Faraday's constant ($Cmol^{-1}$),

V = the volume of the electrolyte (m^3).

The following relationship is also considered accordingly (equation 8) (Comninellis and Pulgarin 1991):

$$ICE_{COD} = [(COD_t - COD_{t+\Delta t}) / 8I\Delta t]. FV, \qquad (8)$$

$(COD)_t$ and $(COD)_{t+\Delta t}$ = the chemical oxygen demands at times t and t+Δt (g_{O2} dm^{-3}),

I = the current (A),

F = Faraday's constant (96,500C mol^{-1}),

V = the volume of the electrolyte (dm^3).

2. Electrochemical Oxidation Index (EOI)

The average value of the current efficiency during the oxidation is so-called the EOI. It is determined from the instantaneous current efficiency (ICE) using the following relationship (equation 9a) (Comninellis and Pulgarin, 1991):

$$EOI = \int ICE dt / \tau \qquad (9a)$$

τ = the time at which the instantaneous current efficiency equals zero.

The average current efficiency at time t is given by following equeation (equation 9b) (Brillas et al., 2009):

$$CE (\%) = [(\Delta COD)_t FV/8It].100 \qquad (9b)$$

$(\Delta COD)_t$ = the experimental decay of the chemical oxygen demand (COD) (gO_2 L^{-1}) at time t (s).

The current efficiency for H_2O_2 accumulation is determind using the following equeation (equation 9c) (Brillas et al., 2009):

$$CE (\%) = [(nFc(H_2O_2)V)/(1000M(H_2O_2)Q)].100 \qquad (9c)$$

n = the ctoichiometric number of electrons transferred in the reaction $O_{2(g)}+2H^+ +2e^- \rightarrow H_2O_2$,

F = the Faraday's constant,
c (H_2O_2) = the accumulated H_2O_2 concentration (mg L^{-1}),
V = the treated solution's volume (L),
1000 = a conversion factor,
M (H_2O_2) = the molecular weight of H_2O_2 (34 g mol^{-1}),
Q = the charge consumed during the electrolysis.

If there are necessary data related to total organic carbon's concentration (TOC) for a single contaminant available, the mineralization current efficiency (MCE) for electrolyzed solutions at time t (h) can be calculated as follows (Brillas et al., 2009):

$MCE (\%) = [(\Delta TOC)_t nFV/4.32x10^7 m I t].100$ (9d)
$(\Delta TOC)_t$ = the TOC decay (mg carbon L^{-1}),
n = the number of electrons exchanged in the organic's mineralization process,
m = the number of carbon atoms of the considered molecule,
4.32×10^7 = the conversion factor for units homogenization (= 3600 s h^{-1} x 12000 mg carbon mol^{-1}).

3. Electrochemical Oxygen Demand (EOD)

EOD is determined employing the following relation (equation 10) (Comninellis and Pulgarin 1991):

$$EOD\ (g_{O2}\ (g_{organic})^{-1}) = 8(EOI \times I \times \tau) / F[organic] \quad (10)$$

[organic] = the amount of the organic compound in the electrolyte (g).

4. General Current Efficiency (GCE)

General current efficiency for the anodic oxidation can be calculated from the available chemical oxygen demand (COD) data employing the following relationship (equation 11) (Martinez-Huitle et al., 2004):

$$GCE = F((COD_0 - COD_t) / 8It) \quad (11)$$

COD_0 and COD_t = the chemical oxygen demands (gdm^{-3}) at time t=0 and t,
I = the current (A)
F = the Faraday's constant,
V = the electrolyte's volume (dm^3),
8 = the oxygen equivalent mass (geq^{-1}).

5. Space – Time Yield (YST)

The overall capacity of the electrochemical oxidation, i.e., in terms of space-time yield can be determined employing the following relationship (equation 12) (Chen 2004):

$$Y_{ST} = ((a \text{ x } i \text{ x CE x M}) / nF).\ 3{,}600 \quad (12)$$

a = the specific electrode area (m^{-1}),
i = the current density (Am^{-2}),
CE = the current efficiency,
n = the number of the involved electrons in the reaction,
F = the Faraday's constant,
M = the molar mass ($gmol^{-1}$).

6. Energy-related Parameters

These parameters are basic figures according to economic-operational purposes. Energy consumption per unit volume, unit COD mass, and unit TOC mass at I=const. can be calculated using the following equeations (equations 13a,b,c) (Brillas et al., 2009):

$$\text{Energy consumption expressed in kWh } m^{-3} = E_{cell}It/V \quad (13a)$$

$$\text{Energy consumption expressed in kWh (g COD)}^{-1} = E_{cell}It/(\Delta COD)_t V \quad (13b)$$

$$\text{Energy consumption expressed in kWh (g TOC)}^{-1}) = E_{cell}It/(\Delta TOC)_t V \quad (13c)$$

E_{cell} = the average cell voltage (V),
t = the electrolysis time (h),
$(\Delta COD)_t$ and $(\Delta TOC)_t$ = the decays in COD (mg O_2 L^{-1}) and TOC (mg carbon L^{-1}).

Anode materials are classified related to their oxidation power in acid media, i.e., the higher the oxygen overvoltage, the higher its oxidation power. The specific character of the high oxidation power anode is a considerably limited electrode-hydroxyl radical interaction. Consequently, a low electrical activity resulted for the oxygen evolution reaction (high overvoltage anode) and a high chemical reactivity resulted for organic compounds oxidation (high current efficiency) are distinguishable (Kapalka et al., 2010). Anodes such as carbon, graphite, IrO_2, RuO_2, and platinum are effective catalysts for the oxygen evolution, i.e., low oxygen evolution overpotential, partial oxidation of the organics. Anodes such as antimony-doped tin dioxide, lead dioxide, and boron-doped diamond are poor catalysts, i.e., high oxygen evolution overpotential, complete oxidation of the organics (Panizza 2010, Quiroz et al., 2014). Thus, the anode materials can be divided into two classes as follows (Comninellis 1994):

Class one anode (active anodes), i.e., low oxygen evolution overpotential, proper electrocatalysts for the oxygen evolution reaction:

- *Carbon and graphite*

  Carbon and graphite have been the main insoluble anodic materials in industrial electrochemical processes for a long time. The low-priced carbon and graphite electrodes have a large surface area and they have been widely used for the removal of organics in electrochemical reactors. However, with these materials the electrooxidation is generally accompanied by surface corrosion, especially at high current densities. Therefore, special graphite anodes with high corrosion resistance combined with high mechanical strength and high purity are developed for industrial processes. Various types of carbon-based electrodes have been employed for the organic compounds treatment (Boudenne et al., 1996; Awad and Abuzaid 1997, 1999, 2000; Polcaro and Palmas 1997; Boudenne and Cerclier 1999; Canizares et al., 1999; Oeguetveren et al., 1999; Polcaro et al., 2000; Piya-areetham et al., 2006; Li et al., 2009; Liu et al., 2012; Zhang et al., 2012).

  Carbon-based materials have also been utilized as cathodes in indirect organics electrolyses producing hydrogen peroxide by oxygen reduction on the cathode surface. Carbon and graphite have shown proper electrochemical behavior related to oxygen reduction, high overpotential for hydrogen evolution, and low catalytic behavior according to hydrogen peroxide decay (Do and Chen 1994; Ponce-de-Leon and Pletcher 1995; Martinez and Bahena 2009; Balci et al., 2009; Oturan et al., 2010; Wang and Chu 2011; Plakas et al., 2013; Chu et al., 2013; Ferrag-Siagh et al., 2014).

  Various environmentally friendly carbon-based electrodes have been investigated such as graphite (Sudoh et al., 1986; Do and Chen, 1993), carbon-polytetrafluoroethylene oxygen diffusion (Brillas et al., 1995; Harrington and Pletcher 1999), and three-dimensional electrodes such as carbon felt (Oturan et al., 1999; Oturan 2000), activated carbon fiber (Wang et al., 2005), reticulated vitreous carbon (Ponce de Leon and Pletcher 1995; Alvarez-Gallegos and Pletcher 1998, 1999), carbon sponge (Oezcan et al., 2008), carbon nanotubes (Fu et al., 2007), and Fe-modified multiwalled carbon nanotubes (Fu et al., 2007).
- *Platinum-based anodes*

  The platinum electrode is one of the most commonly used anodes because of its strong chemical resistance to corrosion (Soriaga and Hubbard 1982; Lamy et al., 1983; Lamy 1984; Comninellis and Pulgarin 1991; Kotz et al., 1991; Stucki et al., 1991; Gattrell and Kirk, 1993; Foti et al., 1997; Rodgers et al., 1999; Bonfatti et al., 1999; Socha et al., 2005; Oturan et al., 2009; Martinez and Bahena 2009; Balci et al., 2009; Panizza 2010; Feng et al., 2014; Ferrag-Siagh et al., 2014; He et al., 2014).
- *Iridium-based oxides*

  The Iridium dioxide anodes (IrO_2) stability, i.e., service life is about 20 times higher than that of the equivalent Ruthenium dioxide (RuO_2) anodes as oxygen evolution electrodes (Alves et al., 1998; Guo et al., 2010).

  Many studies related to the organics oxidation with Ti/IrO_2 electrodes have also been carried out (Pulgarin et al., 1994; Foti et al., 1997; Simond et al., 1997; Foti et al., 1999; Panizza 2010). Several authors have reported that DSA®-type anodes (Dimensionally stable anodes) coated with a layer of IrO_2, RuO_2 and other oxides can be employed for organics destruction by indirect electrolysis producing in situ active chlorine by the existing chloride ions oxidation (Comninellis and Nerini 1995; Do and Yeh 1995; Do et al., 1997; Szpyrkowicz et al., 2000; Panizza and Cerisola 2003; Panizza 2010).

The ternary oxide Ti/IrO_2-Sb_2O_5-SnO_2 electrodes are more stable and effective than IrO_2 (Chen et al., 2001; Chen et al., 2002; Chen and Chen 2005; Guo et al., 2010).

- *Ruthenium-based oxides*

Ruthenium is a component of mixed-metal oxide anodes utilized for electrolytic cells for chemical processes (Cardarelli 2008). RuO_2-based materials have been investigated as electrode materials for various gas evolution electrodes (Sugimoto et al., 2006). The dimensionally stable anodes (DSA®) consist of metal-oxide such as ruthenium-based oxide or mixed metal-oxide coatings on titanium. The dimensionally stable anodes coated with a thin layer of ruthenium-based oxides have been investigated for applications in the organics oxidation process (Do and Yeh 1995; Do et al., 1997; Chandler et al., 1997; Houk et al., 1998; Johnson et al., 1999; Zanta et al., 2000; Szpyrkowicz et al., 2000; Bock and Mac Dougall 2000; Lanza and Bertazzoli 2002; Panizza and Cerisola 2003; Malpass et al., 2006; Costa et al., 2008; Tran et al., 2009; Muhammad et al., 2012; Goncalves et al., 2012; Pipi et al., 2013).

Class two anodes (nonactive anodes) have high oxygen evolution overpotential and are improper electrocatalysts for the oxygen evolution reaction:

- *Tin dioxide, Antimony-doped tin dioxide*

The performance of tin dioxide anodes with high oxygen evolution overpotential has been investigated according to its application for electrochemical oxidation of organic contaminants contained in wastewaters (Koetz et al., 1991; Stucki et al., 1991; Comninellis and Pulgarin 1993; Lipp and Pletcher 1997; Correa-Lozano et al., 1997). Antimony-doped tin dioxide anodes supported on titanium (Ti/SnO_2-Sb) possess a considerable oxygen evolution potential and are considerably effective for the electrochemical oxidation of organic compounds in wastewater treatment. However, its use is limited by its low stability, i.e., short service life. The use of platinum or ruthenium in the oxide layer increases the electrode stability or the electrolytic activity for the oxygen evolution reaction. (Stucki et al., 1991; Comninellis and Pulgarin 1993; Pulgarin et al., 1994; Cossu et al., 1998; Vicent et al., 1998; Bock and Mac Dougall 1999; Polcaro et al., 1999; Grimm et al., 2000; Montilla et al., 2004a,b,c; Montilla et al., 2005; Li et al., 2005; Costa et al., 2008; Berenguer et al., 2009; Cui et al., 2009; Esquivel et al., 2009 ; Costa et al., 2010 ; Vargas et al., 2010; Koivula 2010; Debiemme-Chouvy et al., 2011; Cui et al., 2012; Li et al., 2013; Xu et al., 2013; Duan et al., 2014).

- *Lead dioxide*

Anodes with proper activity and stability have been widely investigated according to electrochemical oxidation for organics removal in wastewater treatment. Lead dioxide (PbO_2) anodes have been employed in this regard because of their suitable conductivity and large overpotential for oxygen evolution in acidic media, making the hydroxyl radicals generation possible during water discharge. The possible release of toxic ions is the main concern accordingly (Cossu et al., 1998; Saracco et al., 2000; Gherardini et al., 2001; Keech and Bunce 2003; Zhou et al.,

2005; Canizares et al., 2005; Samet et al., 2006; Hamni et al., 2009; Panizza 2010, Panizza and Cerisola 2010; Zhang et al., 2014).

The improvement of electrochemical properties, activity, stability and performance has been investigated employing novel PbO_2 electrodes (Zhou et al., 2005; Zhou and He 2008; Yang et al., 2009; Song et al., 2010; Wang et al., 2010; Zhao et al., 2010; Aquino et al., 2011; Chen et al., 2013; Yao et al., 2013; Hu et al., 2014) and PbO_2 doped or coated anodes (Ho et al., 2007; Andrade et al., 2008; Liu et al., 2008; Liu et al., 2011; Subba Rao and Venkatarangaiah 2014).

- *Boron-doped diamond (BDD)*

  Boron-doped diamond electrodes have several important characteristics including inert surface with low adsorption properties, considerable corrosion stability, and high oxygen evolution overpotential. During electrolysis in the region of water discharge, a boron-doped diamond anode makes the generation of a large quantity of the •OH possible with low surface adsorption. Therefore, it is highly active in organics oxidation process (Gandini et al., 1999; Foti et al., 1999; Perret et al., 1999; Gandini et al., 2000; Panizza et al., 2001; Gherardini et al., 2001; Iniesta et al., 2001; Rodrigo et al., 2001; Bellagamba et al., 2002; Troster et al., 2002; Montilla et al., 2002; Canizares et al., 2002; Chen et al., 2003; Kraft et al., 2003; Lissens et al., 2003; Hattori et al., 2003; Canizares et al., 2003; Rychen et al., 2003; Polcaro et al., 2003; Brillas et al., 2004; Fernandes et al., 2004; Morao et al., 2004; Martinez-Huitle et al., 2004; Canizares et al., 2004; Comninellis et al., 2005; Panizza and Cerisola 2005; Polcaro et al., 2005; Panizza et al., 2005; Panizza and Cerisola 2006; Panizza 2010; Ochiai et al., 2011; El-Ghenymy et al., 2012; Garcia-Segura et al., 2012; Wang et al., 2012; Feng et al., 2014; Hernandez et al., 2014).

VARIOUS ELECTROCHEMICAL ADVANCED OXIDATION PROCESSES

Over the last decade, electrochemical advanced oxidation processes have experienced a significant development showing great effectiveness for the decontamination of polluted water and wastewater. The electro-Fenton process was the first method proposed accordingly. Thereafter, various related processes have been developed and performed.

1. Electro-Fenton (EF) Process

The electro-Fenton process combines the advantages of electrochemical and Fenton processes (Yu et al., 2013). Notable advantages possess this method compared with the chemical Fenton process (Oturan et al., 1999; Oturan 2000; Oturan et al., 2007). The electro-Fenton is one of the powerful and environmentally friendly emerging and promising water and wastewater remediation processes. Electro-Fenton process occurs in different electrolytic reactors such as bubble reactor, filter press reactor, divided double-electrode electrochemical cell, divided three-electrode electrochemical cell and double compartment cell (Nidheesh and Gandhimathi 2012).

The electro-Fenton process is based on the continuous hydrogen peroxide formation at a proper cathode fed with oxygen or air, and the addition of an iron catalyst (Fe^{2+}, Fe^{3+}, or iron oxides) to the treated solution at the same time to generate hydroxyl radical (oxidant $^{\bullet}OH$) at the bulk according to the Fenton's reaction as follows (Pignatello et al., 2006):

$$Fe + H_2O_2 \rightarrow Fe^{3+} + {}^{\bullet}OH + OH^- \qquad (14)$$

The process occurs in undivided cells under relatively complex conditions because of the simultaneous contaminants decomposition with reactive oxygen species, i.e., particularly the hydroxyl radical generated at the anode. The oxidation of cathodically generated hydrogen peroxide at the anode of the undivided cells according to following reactions (equations 15 and 16) (Brillas et al., 1995) causes a considerable loss of the hydrogen peroxide's accumulation in contrast with divided cells. An advantage of the undivided cells performance is the necessary lower cell voltage for the electrolysis process. (Brillas et al., 1996, 2000; Boye et al., 2002; Kornienko et al., 2004; Wang et al., 2005; Wang et al., 2005; Badellino et al., 2006; Fu et al., 2007; Flox et al., 2007; Wang and Wang 2007; Panizza and Cerisola 2008; Da Pozzo et al., 2008; Oezcan et al., 2008; Brillas et al., 2009).

$$H_2O_2 \rightarrow HO^{\bullet}_2 + H^+ + e^- \qquad (15)$$

$$HO^{\bullet}_2 \rightarrow O_{2(g)} + H^+ + e^- \qquad (16)$$

The organic compounds decomposition occurs in an undivided cell via heterogeneously generated hydroxyl radical M ($^{\bullet}OH$). It is generated at a high oxygen overvoltage anode M from water oxidation according to the reaction 17 (Panizza et al., 2001; Panizza and Cerisola 2005; Flox et al., 2006):

$$M + H_2O \rightarrow M\,({}^{\bullet}OH) + H^+ + e^- \qquad (17)$$

The so-called paired electrocatalysis process occurs employing high oxidation power anodes to minimize the extent of oxygen evolution from the following reaction (the reaction18) (Brillas et al., 2009):

$$2M\,({}^{\bullet}OH) \rightarrow 2M + O_2 + 2H^+ + 2e^- \qquad (18)$$

According to the oxidative action of M ($^{\bullet}OH$), the boron-doped diamond anode is more efficient than platinum, tin dioxide, lead dioxide, Iridium dioxide, and ruthenium dioxide (Panizza et al., 2001; Panizza and Cerisola 2005; Weiss et al., 2008).

It must be notified that by the electro-Fenton decomposition of organic compounds in an undivided cell, parallel destruction with weaker oxidizing species such as $HO_2^{\bullet}$, H_2O_2, O_3, ferrate ion and other hypervalent iron species, $S_2O_8^{2-}$ ion, Cl_2, etc. is also possible. The effect of oxidizing species is not significant when a carbon-felt cathode is employed. In contrast, the influence of the anode is very considerable when a gas diffusion electrode is used as cathode, because of its poor ability related to the Fe^{2+} regeneration. The combination of both anode and cathode reactions for water treatment has also been proposed for divided cells. Various two- and three electrode divided cells have been used for the hydrogen peroxide electrogeneration. These systems with two separate solutions, i.e., catholyte and anolyte are performed under potentiostatic or galvanostatic conditions (Sudoh et al., 1985; Sudoh et al., 1986; Do and Chen 1993, 1994; Do and Yeh 1996, 1998; Do et al., 1997; Alvarez-Gallegos and Pletcher 1998; Kornienko et al., 2004; Wang et al., 2005; Da Pozzo, 2005; Boye et al., 2006; Wang and Wang 2007; Liu et al., 2007; Agladze et al. 2007; Panizza and Cerisola 2008; Brillas et al., 2009). (Brillas et al., 2009).

In the electro-Fenton process and several related methods, the cathodically reduction of soluble Fe^{3+} to Fe^{2+} occurs as follows (the reaction 19) (electrochemical catalysis) (Aaron et al., 2001):

$$Fe^{3+} + e^{-} \rightarrow Fe^{2+} \quad (19)$$

The fast regeneration of Fe^{2+} accelerates the generation of the $^{\bullet}OH$ resulted from Fenton's reaction 14. Therefore, both Fe^{3+} and Fe^{2+} ions can be used as alternative iron sources with three-dimensional carbonaceous materials for the continual formation of a high amount of $^{\bullet}OH$ (Oturan et al., 2008; Brillas et al., 2009). In this regard, the possibility of the alternative use of Cu^{2+}/Cu^{+} (Oturan and Oturan 2005; Pimentel et al., 2008), Co^{3+}/Co^{2+} (Chen et al., 2003; Pimentel et al., 2008), and Mn^{3+}/Mn^{2+} (Pimentel et al., 2008) couples instead of the Fe^{3+}/Fe^{2+} pair has been investigated.

The organics degradation rate in the electro-Fenton process depends on operation parameters such as oxygen feeding, stirring rate or liquid flow rate, temperature, solution pH, electrolyte composition, applied potential or current, iron catalyst and initial contaminant concentration (Brillas et al., 2009).

In the last decades, various combined electrochemical methods based on Fenton's reaction have been investigated to ameliorate the electro-Fenton process efficiency as follows:

- Peroxi-Coagulation process,
- Photoelectro-Fenton process,
- Photoperoxi-Coagulation process,
- Photoelectrocatalysis process,
- Photoelectrochemical electro-Fenton process,
- Solar Photoelectro-Fenton process,
- Sonoelectro-Fenton process,
- Cathodic Generation of Fe^{2+}.

In the recent decade, several novel applications of the electro-Fenton process have been investigated and carried out. Some of them are as follows:

Fe_3O_4 magnetic nanoparticles were employed in an undivided electrochemical reactor with an activated carbon felt cathode and a platinum anode for the C.I. Reactive Blue 19 degradation purpose (He et al., 2014).

The electrogeneration of hydrogen peroxide and decomposition of an organic contaminant (methyl orange) by a novel electro-Fenton process has been investigated utilizing dual rotating graphite felt disks to supply oxygen (Yu et al., 2014).

The treatment of Basic Red 29 dye solution was investigated employing hybrid iron-aluminum electrodes by electrocoagulation and electro-Fenton methods. The electro-Fenton system possessed higher effectiveness. However, both systems were suitable processes for treating textile wastewater including dyes (Yavuz et al., 2014).

The electro-Fenton method was investigated for treating phenanthrene contaminated sediment. After one day, under optimized condition, i.e., the optimized surfactant mixture, removal efficiency was near 100%. (Iglesias et al., 2014).

The degradation of the nonsteroidal anti-inflammatory drug Ketoprofen in tap water was investigated utilizing electro-Fenton and anodic oxidation processes using platinum and boron-doped diamond anodes and carbon felt cathode. The fast degradation and complete mineralization was achieved by doron-doped diamond/carbon felt, platinum/carbon felt, and anodic oxidation using boron-doped diamond anode (Feng et al., 2014).

A bio-electro-Fenton process was carried out for arsenite As (III) oxidation in a dual-chamber microbial fuel cell (MFC) and neutral pH conditions. The hydrogen peroxid and Fe^{2+} reagents were generated in situ by microbial-driven electro-reduction of oxygen and γ-FeOOH (Wang et al., 2014).

The feasibility of an electro-Fenton process to treat the non-biodegradable antibiotic tylosin was investigated utilizing a discontinuous electrochemical cell with divided cathodic and anodic compartments.The total degradation of tylosin was resulted after 15 minutes employing a carbon felt cathode and a platinum anode; However, six hours was needed to achieve 88-96% oxidation and mineralization yields (Ferrag-Siagh et al. 2014).

A novel electro-Fenton filter was investigated for toxic organics degradation, e.g., pesticides and pharmaceuticals from water. It was comprised of a stack of carbon anodic and cathodic electrode pairs for operation in continuous mode (Plakas et al., 2013).

The application of the electro-Fenton process for coliform removal from leachate was investigated. The study focused on leachates with low and high coliform content. The Electro-Fenton process considered possessed high removal efficiency, especially in removing high concentrations of pathogenic microorganisms from semi-aerobic stabilized leachate (Hamidi et al., 2013).

The performance the electro-Fenton process using Fe alginate gel beads for the treatment of a wastewater contaminated with synthetic dyes (Reactive Black5) was investigated (Iglesias et al., 2013).

A novel approach was proposed to regulate the effective Fe^{2+} generation from an iron cathode in a three-electrode system in addition to hydrogen peroxide production. The Fe^{2+} was used for the Pd-catalytic transformation of methyl *tert*-butyl ether in water (Liao et al., 2013).

A series of ferrite-carbon aerogel monoliths with different iron/carbon ratios was synthesized as high catalytic cathode of an electro-Fenton system. The ferrite-carbon aerogel possessed good conductivity, high catalytic efficiency, and considerable stability. A 98% TOC elimination was realized after four hours (Wang et al., 2013).

The optimal reuse of iron-rich sludge for an electro-Fenton type process employing sequencing batch mode and separation batch mode reuse models was studied. The Iron from the sludge investigated could be reused for the electrochemical process when the separation batch mode was employed, with complete iron recovery and without decline in current efficiency (Kishimoto et al., 2013).

A trickle bed cathode was performed by coating a layer composed of carbon black and polytetrafluoroethylene onto graphite chips to avoid electrolyte leakage and gas bubbles in the electro-Fenton system utilizing a gas diffusion cathode (Lei et al., 2013).

The degradation of *m*-cresol solution was investigated employing an electro-Fenton system with Ti/SnO_2-Sb_2O_5-IrO_2 anode for anodic oxidation and porous carbon-nanotube cathode for hydrogen peroxide generation along with Fe^{3+} reduction. The porous cathode made of graphite, carbon nanotube and polytetrafluoroethene possessed high catalytic activity.The system was able to considerably enhance the biodegradability of *m*-cresol solution (Chu et al., 2013).

UV-A, UV-C, and visible-light assisted photo-Fenton-like systems for degrading Bisphenol A were investigated in pure water and raw freshwater. Total degradation of the Bisphenol A accompanied with considerable removal of the DOC was achieved for all studied processes (Molkenthin et al., 2013).

The mineralization of sulfanilic acid was investigated by electro-Fenton and Photoelectro-Fenton reactions with UV-A light utilizing an undivided electrochemical cell employing a boron-doped diamond anode and an air diffusion cathode (El-Ghenymy et al., 2012).

The mineralization of the antimicrobial agent flumequine was investigated by electrochemical advanced oxidation processes, i.e., electro-Fenton and photoelectro-Fenton system with UV-A light employing a boron-doped diamond anode and an air-diffusion cathode at constant current. The photoelectron-Fenton process was more powerful in this regard, i.e., total mineralization with 94-96% total organic carbon removal (Garcia-Segura et al., 2012).

The microwave enhanced continuous flow electro-Fenton system for the treatment of azo-dye wastewater employing boron-doped diamond anode was performed under atmospheric pressure. This study improved the theory of catalytic oxidation combination technology and presented a novel proposal for the removal of concentrated refractory organic contaminants (Wang et al., 2012).

A novel three-dimensional electro-Fenton system was investigated for wastewater treatment using foam nickel, activated carbon fiber, and Ti/RuO_2-IrO_2 as the particle electrodes, the cathode, and the anode respectively. This system possessed a noticeable high rhodamine B removal efficiency, i.e., equals 99% (Liu et al., 2012).

The improvement in electro-Fenton method's ability for the wastewater treatment contaminated with synthetic dyes was investigated by means of the central composite face-centered experimental design matrix and the response surface methodology (Rosales et al., 2012).

Activated carbon-supported nano-FeOOH catalysts were prepared for efficient adsorption and combined heterogeneous/homogeneous Fenton oxidation of the amaranth. The heterogeneous catalyst showed considerable adsorbability. The electrochemically assisted amaranth oxidation process was occured via the heterogeneous Fenton reaction on catalyst/solution interface, and the homogeneous Fenton reaction in bulk solution (Zhang et al., 2012).

A novel high effective electro-Fenton process was developed for wastewater treatment utilizing a modified divided electrolytic system. Hydrogen peroxide was generated in situ from electrogenerated hydrogen and oxygen employing the Pd/C catalyst. The performance of the novel process was assessed by Rhodamine B degradation in an aqueous solution (Yuan et al., 2011).

A novel enhanced electrochemical advanced oxidation process for the degradation of organic contaminants in aqueous solution was investigated employing an iron sheet as sacrificial anode and a graphite bar as cathode. The coupled process of Fe(II)/Oxone and electrochemical treatment, i.e., electro-Fe(II)/Oxone was evaluated related to 2,4,5-Trichlorophenoxyacetic acid degradation in aqueous solution. The degradation efficiency equaled over 90% decay within 10 min (Wang and Chu 2011).

The degradation of the insecticide metomyl (Lanate 20L) by an electro-Fenton process using a carbon felt cathode and iron ions as catalyst was investigated and compared with transition metal ions, i.e., cobalt, silver, and copper as catalyst to generate hydroxyl radicals. The effect of the catalyst was investigated and optimized accordingly (Oturan et al., 2010).

A novel approach utilizing the biological electrons produced from a microbial fuel cell to power an electro-Fenton process for wastewater treatment at neutral pH, so-called bioelectro-Fenton process was investigated (Feng et al., 2010).

A novel sandwich film cathode based electro-Fenton system was investigated. Two pieces of Fe-chitosan/nickel were used to fasten one piece of activated carbon fiber to gain a Fe-CHI/Ni|activated carbon fiber|Fe-CHI/Ni cathode. This novel system could effectively decompose the rhodamine B with in situ generating both hydrogen peroxide and iron ions. This low-cost cathode possessed a high stability and was reuseable without decreasing catalytic activity (Fan et al., 2010).

A modified electro-Fenton-like process for the glyphosate aqueous solutions degradation was investigated employing Mn^{2+} and other metal ions as catalysts. Bulk electrolyses with a carbon-felt cathode and platinum anode were used in an undivided cell under galvanostatic conditions. The herbicide was totally destroyed in all cases (Balci et al., 2009).

The degradation of low concentration of chlorbromuron in aqueous solutions was investigated by electro-Fenton process comprised of three-electrode divided and undivided cell employing reticulated vitreous carbon cathode and platinum anode. The electro-Fenton process was also studied in a two-electrode undivided cell using a sacrificial iron anode. The TOC removal equaled over 98% (Martinez and Bahena 2009).

A heterogeneous electro-Fenton system with a novel $Fe@Fe_2O_3$/active carbon fiber oxygen diffusion cathode was investigated for the degradation of the dye contaminant rhodamine B in water. This novel system possessed a notable high activity at neutral pH (Li et al., 2009).

The Kinetics of the degradation of several chlorophenols by the electro-Fenton process was investigated employing carbon felt cathode and platinum anode. The results showed that more chlorinated phenols are more difficult to mineralize (Oturan et al., 2009).

The electrocatalytic activities of the bare graphite and the anthraquinonedisulfonate /polypyrrole composite film modified graphite cathodes according to oxygen reduction and Fe^{2+}regeneration, and also the electro-Fenton degradation of the amaranth azo-dye using the bare graphite and the anthraquinonedisulfonate/polypyrrole/ graphite as cathode and Fe^{2+} as catalyst was investigated (Zhang et al., 2008).

Core-shell $Fe@Fe_2O_3$ nanowires and multiwall carbon nanotubes were combined with polytetrafluorethylene and utilized as an oxygen gas diffusion electrode in an electro-Fenton system. The degradation of rhodamine B equaled 91.5% in two hours at neutral pH (Ai et al., 2007).

An integrated technology, i.e., an electrochemical pre-treatment step of the wastewater using the electro-Fenton reaction followed by an anaerobic bio-treatment for the treatment of the recalcitrant contaminants of olive mill wastewater was investigated. The electro-Fenton process removed 65.8% of the total polyphenolic compounds and subsequently decreased the wastewater's toxicity from 100% to 66.9%. The use of electro-coagulation as post-treatment method detoxified the anaerobic effluent (Khoufi et al., 2006).

The viability of the degradation of aqueous solutions of aniline, nitrobenzene, and 4-chlorophenol by the photoelectro-Fenton process employing a gas diffusion cathode in a pilot reactor using solar and UVA irradiation was investigated. An intensity of 20A was applied in a flow cell comprising a Ti/Pt anode and a carbon-PTFE cathode. After electrolysis, samples of the effluent were exposed to sunlight, and almost complete mineralization was reached after ca. 50 min (Casado et al., 2005).

2. Fered – Fenton Process

The Fered-Fenton process (EF-Fere) occurs in an undivided electrolytic cell and is comprised of adding Fe^{2+} or Fe^{3+} to the wastewater in addition to the simultaneously continuous injection of hydrogen peroxide during the treatment. Fe^{3+} is then cathodically reduced to Fe^{2+} (Eq.19). Fe^{2+} reacts with the hydrogen peroxide (H_2O_2) to produce $^{\bullet}OH$ via Fenton's reaction 14. H_2 results from water reduced at the cathode (the reaction 20). The H_2 oxidation to $^{\bullet}OH$ occurs at the anode surface (the reaction 17) (Brillas et al., 2009):

$$2H_2O + 2e^- \rightarrow H_{2(g)} + 2OH^- \qquad (20)$$

The related operation parameters, i.e., the concentration of iron ions, pH and cathodic current density must be optimized. In contrast to electro-Fenton and combined electro-Fenton systems, the Fered-Fenton method uses a higher content of iron ions, which act as reactants rather than catalysts.

The photoassisted Fered-Fenton degradation of organics with simultaneous UV-A illumination of the wastewater is also an effective method. The UV-A light causes a fast photodecomposition of the Fe^{3+}complexes with final carboxylic acids from reaction $Fe(OOCR)^{2+} + h\nu \rightarrow Fe^{2+} + CO_2 + R^{\bullet}$.

In the recent years, only a few related studies have been applied as follows:

Waste-activated sludge stabilization utilizing Fered-Fenton process was investigated considering organics degradation efficiency. The equipment consists of two graphite anodes and two graphite cathodes. The correlations of the investigated parameters were determined by employing the Kohonen self-organizing feature maps (KSOFM). The multi-layer perceptron (MLP) method was used for a comprehensive evaluation of the parameters correlation and prediction aims. The results showed that the Fered-Fenton reactor could be an efficient process for reducing the waste-activated sludge organic load considerably (Badalians Gholikandi et al., 2014).

A factorial design was carried out to evaluate the leachate treatment application utilizing photoassisted Fered-Fenton method. The correlation between COD removal and the significant independent variables was established employing an experimental design. The hydrogen peroxide and Fe^{2+} concentrations, initial pH, and the interaction effect between current density and initial pH were all significant factors (Wu et al., 2014).

The citrate remediation was carried out by utilizing the Fered-Fenton technology. Thereafter, the feed nickel ions removal was performed by employing the alkaline precipitation process. The nickel citrate was oxidized in an electrochemical reactor that consists of four anodic rod (Ti/RuO_2-IrO_2) and cathodic net (stainless steel) pairs. The high COD removal equaled 93% accordingly (Shih et al., 2013).

The affecting operation conditions of a Fered-Fenton system, i.e., H_2O_2 to Fe^{2+} mole ratio, Fenton's reagent dosage, initial pH, current density, inter-electrode gap and hydrogen peroxide feeding mode were investigated related to the COD removal. One pair of Ti/RuO_2-IrO_2 anode and stainless steel cathode was used accordingly. The organics were degraded into smaller organic products with molecular weight below 4kDa (Zhang et al., 2012).

3. Electrochemical Peroxidation Process (ECP)

Electrochemical peroxidation process or EF-Feox process occurs in a electrolytic cell using a sacrificial iron or steel anode for Fe^{2+} electrogeneration by the reaction $Fe \rightarrow Fe^{2+} + 2e^-$ and an inert cathode for H_2 production from H_2O reduction. Hydrogen peroxide is externally added to the treated solution for the organics oxidation with $^{\bullet}OH$ formed from Fenton's reaction (the reaction 14). In this regard, the anodic Fenton treatment (AFT) has also been considered. The two technologies are differed on the utilization of undivided or divided cells. Consequently, the cathodic regeneration of Fe^{2+} from reaction $Fe^{3+} + e^- \rightarrow Fe^{2+}$ occurs only in the electrochemical peroxidation system. However, in both systems, the controlled Fe^{2+} electrogeneration make the minimization of the extent of the reaction $Fe^{2+} + {}^{\bullet}OH \rightarrow Fe^{3+} + OH^-$ and the accumulation of iron sludge possible. It must be considered that the coagulation with the Fe $(OH)_3$ precipitate depending on pH and the current is an alternative route for the removal of the organic compounds.

The electrochemical peroxidation system was utilized for the synthetic solutions degradation of various herbicides and real wastewaters. The COD removal efficiency of the electrochemical peroxidation system is higher than direct electrocoagulation process. The stepwise dosing of the hydrogen peroxide causes an improvement of the system because of the controlled hydrogenperoxide addition and the minimization of the waste of $^{\bullet}OH$ by reaction $H_2O_2 + {}^{\bullet}OH \rightarrow H_2O + HO_2^{\bullet}$. Another affecting operation factor is the $[Fe^{2+}]/[H_2O_2]$.

Thus, the considerable removal of volatile organic compounds and polychlorinated biphenyls from liquids and sludge is reported as possible using a minimal quantity of iron sludge formed. The application of the electrochemical peroxidation system extended with mild steel electrodes was successfully employed for the effective removal of As, Cr, Pb, Se, and V at pH 3.5-4.6. The photoassisted electrochemical peroxidation systems with UV-A and UV-C light, and the electrochemical peroxidation systems with biological post-treatment are other enhanced methods accordingly (Brillas et al., 2009). In the recent decade, few studies have been reported as follows:

The peroxi-coagulation process the aqueous solutions treatment containing hazardous dye and Rhodamine B was investigated using graphite cathode and iron anode. Various operation parameters, i.e., pH, voltage, electrode area, and different ions were considered accordingly. The experimental results showed that the dye removal efficiency of the pH-regulated peroxi-coagulation system equaled 95% removal after 180 minutes, and the anions such as carbonate, bicarbonate, chloride and sulphate negatively affected the efficiency (Nidheesh and Gandhimathi 2014).

The COD removal efficiency of the pharmaceutical wastewater treatment processes i.e., electro-coagulation, photo-electrocoagulation, peroxi-electrocoagulation and peroxi-photoelectrocoagulation was investigated. The results indicated that under the optimum operating range for each process, the COD removal efficiency of the peroxi-electrocoagulation was the highest (Farhadi et al., 2012).

The decolorization of C.I.Basic Yellow 2 by peroxi-coagulation process employing carbon nanotube-PTFE cathode was investigated. The effect of the four main parameters, i.e., electrolysis time, initial pH, applied current and initial concentration of the dye solution was considered by means of the response surface methodology accordingly. The decolorization process efficiency equaled more than 95% (Zarei et al., 2010).

The degradation efficiency of the peroxi-coagulation process related to 4-Chlorophenoxyacetic acid, 4-chloro-2-methylphenoxyacetic acid, 2,4-dichlorophenoxyacetic acid, 2,4,5-trichlorophenoxyacetic acid, and 3,6-dichloro-2-methoxybenzoic acid was investigated. This electrochemical method yields considerable decontimination of all compounds in acidic aqueous medium. The efficiency of this method decreases with increasing electrolysis time and the current. A faster degradation using the photoperoxi-coagulation process (UV irradiation) of solutions occurs because of the production of more hydroxyl radicals from the photo-Fenton reaction (Brillas et al., 2003; Boye et al., 2003).

The anilin solution's degradation in Na_2SO_4+H_2SO_4 by peroxi-coagulation process at constant pH, temperature and current was investigated utilizing a pilot flow reactor in recirculation mode with a filter-press cell consisting of an anode (Fe, Ti/Pt, DSA®) and an oxygen diffusion cathode. The peroxi-coagulation process efficiency using an Fe anode was more than 95%, since some intermediates coagulate with $Fe(OH)_3$ precipitate formed (Brillas and Casado 2002).

4. Plasma Assisted Fenton Process

The nonthermal plasma oxidation method produces reactive species such as high-energy electrons, $^{\bullet}OH$, $H^{\bullet}$, $O^{\bullet}$, and $O_2^{\bullet-}$, as well as long-lived active molecules such as O_3, H_2O_2, excited-state neutral molecules, and ionic species. The addition of Fe^{2+} or Fe^{3+} considerably accelerates the degradation rate of organics because of the extra production of $^{\bullet}OH$ with generated H_2O_2 from Fenton's reaction (the reaction 14). However, this method possess excessive energy requirement (Abdelmalek et al., 2008; Zhang et al., 2008; Brillas et al., 2009; Mozgina et al., 2009; Magureanu et al., 2010; Magureanu et al., 2011; Schmidt-Szalowski et al., 2011; Vandenbroucke et al., 2011; Wang and Chen 2011; Karuppiah et al., 2012).

REFERENCES

Aaron, J.J., Oturan, M.A. (2001) New Photochemical and Electrochemical Methods for the Degradation of Pesticides in Aqueous media. Environmental Applications, *Turkish Journal of* Chemistry 25(4), Tuebitak, 509-520.

Abdelmalek, F., Torres, R.A., Combet, E., Petrier, C., Pulgarin, C. & Addou, A. (2008) Gliding Arc Discharge (GAD) assisted catalytic degradation of bisphenol A in solution with ferrous ions, *Separation and Purification Technology* 63(1), Elsevier, 30-37.

Abdessalem, A.K., Oturan, N., Bellakhal, N., Dachraoui, M. & Oturan, M.A. (2008) Experimental design methodology applied to electro-Fenton treatment for degradation of herbicide chlortoluron, *Applied Catalysis* B: Environmental 78(3), 334-341.

Abeish, A.M., Ang, M. & Znad, H. (2014) Enhanced Solar-Photocatalytic Degradation of Combined Chlorophenols Using Ferric Ions and Hydrogen Peroxide, *Journal of Industrial & Engineering Chemistry Research* 53(26), American Chemical Society (ACS), 10583-10589.

Agladze, G.R., Tsurtsumia, G.S., Jung, B.I., Kim, J.S. & Gorelishvili, G. (2007) Comparative study of hydrogen peroxide electro-generation on gas-diffusion electrodes in undivided and membrane cells, *Journal of Applied Electrochemistry* 37(3), Springer, 375-383.

Ahmed, B., Limem, E., Abdel-Wahab, A. & Nasr, B. (2011) Photo-Fenton Treatment of Actual Agro-Industrial Wastewaters, *Journal of Industrial & Engineering Chemistry Research* 50(11), American Chemical Society (ACS), 6673-6680.

Ai, Z., Mei, T., Liu, J., Li, J., Jia, F., Zhang, L. & Qiu, J. (2007) Fe@Fe_2O_3 Core-Shell Nanowires as an Iron Reagent. 3. Their Combination with CNTs as an Effective Oxygen-Fed Gas Diffusion Electrode in a Neutral Electro-Fenton System, *the Journal of Physical Chemistry* C 111(40), American Chemical Society (ACS), 14799-14803.

Altin, A. (2008) An alternative type of photoelectron-Fenton process for the treatment of landfill leachate, *Separation and Purification Technology* 61(3), Elsevier, 391-397.

Alvarez-Gallegos, A., Pletcher, D. (1998) The removal of low level organics via hydrogen peroxide formed in a reticulated vitreous carbon cathode cell, Part 1: The electrosynthesis of hydrogen peroxide in aqueous acidic solutions, *Electrochimica Acta,* 44(5), the International Society of Electrochemistry (ISE), Elsevier, 853-861.

Alverez-Gallegos, A., Pletcher, D. (1999) The removal of low level organics via hydrogen peroxide formed in a reticulated vitreous carbon cathode cell. Part 2: The removal of phenols and related compounds from aqueous effluents, *Electrochimica Acta* 44(14), the International Society of Electrochemistry (ISE), Elsevier, 2483-2492.

Alves, V.A., Silva, L.A.D., Oliveira, E.D. & Boodts, J.F.C. (1998) Investigation under conditions of accelerated anodic corrosion of the effect of TiO_2 substitution by CeO_2 on the stability of Ir-based ceramic coatings, *Materials* Science Forum 289-292, TransTechPublications INC., 655-666.

Andrade, L.S., Rocha-Filha, R.C., Bocchi, N., Biaggio, S.R., Iniesta, J., Garcia-Garcia, V. & Montiel, V. (2008) Degradation of phenol using Co- and Co,F-doped PbO_2 anodes in electrochemical filter-press cells, *Journal of Hazardous Materials* 153(1-2), Elsevier, 252-260.

Aquino, J.M., Pereira, G.F., Rocha-Filho, R.C., Bocchi, N. & Biaggio, S.R. (2011) Electrochemical degradation of a real textile effluent using boron-doped diamond or β-PbO_2 as anode, *Journal of Hazardous Materials* 192(3), Elsevier, 1275-1282.

Awad, Y.M., Abuzaid, N.S. (1997) Electrochemical treatment of phenolic wastewater: Efficiency, design considerations and economic evaluation, Journal of Environmental Science and Health, Part A, *Toxic/Hazardous Substances and Environmental Engineering 32*, Taylor & Francis, 1393-1414.

Awad, Y.M., Abuzaid, N.S. (1999) Electrochemical oxidation of phenol using graphite anodes, *Separation Science and Technology* 34(4), Taylor & Francis, 699-708.

Awad, Y.M., Abuzaid, N.S. (2000) The influence of residence time on the anodic oxidation of phenol, *Separation and Purification Technology* 18(3), Elsevier, 227-236.

Bacardit, J., Sans, C., Seminago, R. & Esplugas, S. (2010) Characterization and Control Strategies of an Integrated Chemical-Biological System for the Remediation of Toxic Pollutants in Wastewater: A Case of Study, *Journal of Industrial & Engineering Chemistry Research* 49(15), American Chemical Society (ACS), 6972-6976.

Bach, A., Zelmanov, G. & Semiat, R. (2009) Wastewater mineralization using advanced oxidation process, *Desalination and Water treatment* 6, 152-159.

Badalians Gholikandi G., Masihi, H., Azimipour, M., Abrishami, A. & Mirabi, M. (2014) Optimizing stabilization of waste-activated sludge using Fered-Fenton process and artificial neural network modeling (KSOFM, MLP), *Journal of Environmental Science and Pollution Research* 21(11), Springer, 7177-7186.

Badellino, C., Rodrigues, C.A. & Bertazzoli, R. (2006) Oxidation of pesticides by in situ electrogenerated hydrogen peroxide: Study for the degradation of 2,4-dichlorophenoxyacetic acid, *Journal of Hazardous Materials* 137(2), Elsevier, 856-864.

Balci, B., Oturan, M.A., Oturan, N. & Sires, I. (2009) Decontamination of Aqueous Glyphosate (Aminomethyl) phosphonic Acid and Glufosinate Solutions by Electro-Fenton-Like Process with Mn^{2+} as the Catalyst, *Journal of Agricultural and Food Chemistry 57*(11), American Chemical Society (ACS), 4888-4894.

Bellagamba, R., Michaud, P.A., Comninellis, C. & Vatistas, N. (2002) Electro-combustion of polyacrylates with boron-doped diamond anodes, *Electrochemistry Communications* 4(2), the International Society of Electrochemistry (ISE), Elsevier, 171-176.

Beltran, F.J., Encinar, J.M. & Gonzalez, J.F. (1997) Industrial wastewater advanced oxidation.Part 2. Ozone combined with hydrogen peroxide or UV radiation, *Water Research* 31(10), the International Water Association (IWA), Elsevier, 2415-2428.

Berenguer, R., Quijada, C. & Morallon, E. (2009) Electrochemical characterization of SnO_2 electrodes doped with Ru and Pt, *Electrochimica Acta* 54(22), the International Society of Electrochemistry (ISE), Elsevier, 5230-5238.

Bock, C., Mac Dougall, B. (2000) Anodic oxidation of p-benzoquinone and maleic acid, *Journal of The Electrochemical Society(JES)* 146(8), The society for solid-state and electrochemical science and technology (ECS), 2925-2932.

Bonfatti, F., Ferro, S., Lavezzo, F., Malacarne, M., Lodi, G. & De Battisti, A. (1999) Electrochemical Incineration of Glucose as a Model Organic Substrate, I, Role of the Electrode Material, *Journal of The Electrochemical Society* 146(6), the society for solid-state and electrochemical science and technology (ECS), 2175-2179.

Boudenne, J.L., Cerclier, O. (1999) Performance of carbon black-slurry electrodes for 4-chlorophenol oxidation, *Water Research* 33(2), the International Water Association (IWA), Elsevier, 494-504.

Boudenne, J.L., Cerclier, O., Galea, J. & Vlist, E.V.D. (1996) Electrochemical oxidation of aqueous phenol at a carbon black slurry electrode, *Applied Catalysis A: General* 143(2), Elsevier, 185-202.

Boudreau, J., Bejan, D., Li, S. & Bunce, N.J. (2010) Competition between Electrochemical Advanced Oxidation and Electrochemical Hypochlorination of Sulfamethoxazole at a Boron-Doped Diamond Anode, *Journal of Industrial & Engineering Chemistry Research* 49(6), American Chemical Society (ACS), 2537-2542.

Boye, B., Brillas, E., Buso, A., Farnia, G., Flox, C., Giomo, M. & Sandona, G. (2006) Electrochemical removal of gallic acid from aqueous solutions, *Electrochimica Acta* 52(1), the International Society of Electrochemistry (ISE), Elsevier, 256-262.

Boye, B., Dieng, M.M. & Brillas, E. (2002) Degradation of Herbicide 4-Chlorophenoxyacetic Acid by Advanced Electrochemical Oxidation Methods, *Environmental Science & Technology* 36(13), American Chemical Society (ACS), 3030-3035.

Boye, B., Dieng, M.M. & Brillas, E. (2003) Electrochemical degradation of 2,4,5-trichlorophenoxy-acetic acid in aqueous medium by peroxi-coagulation. Effect of pH and UV light, *Electrochimica Acta* 48(7), the International Society of Electrochemistry (ISE), Elsevier, 781-790.

Brillas, E., Banos, M.A., Camps, S., Arias, C., Cabot, P.L., Garrido, J.A. & Rodríguez, R.M. (2004) Catalytic effect of Fe^{2+}, Cu^{2+} and UVA light on the electrochemical degradation of nitrobenzene using an oxygen-diffusion cathode, *New journal of chemistry* 28(2), 314-322.

Brillas, E., Bastida, R.M., Llosa, E. & Casado, J. (1995) Electrochemical Destruction of Aniline and 4-Chloroaniline for Wastewater Treatment Using a Carbon-PTFE O_2-Fed Cathode, *Journal of The Electrochemical Society* (JES) 142(6), the society for solid-state and electrochemical science and technology, 1733-1741.

Brillas, E., Boye, B., Banos, M.A. & Garrido, J.A. (2003) Electrochemical degradation of chlorophenoxy and chlorobenzoic herbicides in acidic aqueous medium by the peroxi-coagulation method, *Chemosphere* 51(4), Elsevier, 227-235.

Brillas, E., Boye, B., Sires, I., Garrido, J.A., Rodriguez, R.M., Arias, C., Cabot, P.L. & Comninellis, C. (2004) Electrochemical destruction of chlorophenoxy herbicides by anodic oxidation and electro-Fenton using a boron-doped diamond electrode, *Electrochimica Acta* 49(25), the International Society of Electrochemistry (ISE), Elsevier, 4487-4496.

Brillas, E., Cabot, P.L. & Casado, J. (2003) Electrochemical Methods for Degradation of Organic Pollutants in Aqueous Media, In: *Chemical Degradation Methods for Wastes and Pollutants, Environmental and Industrial Applications*, Part 6, Edited by M.A. Tarr, Marcel Dekker Inc., New York-Basel.

Brillas, E., Calpe, J.C., Casado, J. (2000) Mineralization of 2,4-D by advanced electrochemical oxidation processes, *Water Research* 34(8), the International Water Association (IWA), Elsevier, 2253-2262.

Brillas, E., Casado, J. (2002) Aniline degradation by Electro-Fenton® and peroxi-coagulation processes using a flow reactor for wastewater treatment, *Chemosphere* 47(3), Elsevier, 241-248.

Brillas, E., Mur, E. & Casado, J. (1996) Iron (II) Catalysis of the Mineralization of Aniline Using a Carbon-PTFE O_2-Fed Cathode, *The Electrochemical Society* (JES) 143(3), the society for solid-state and electrochemical science and technology, L49-L53.

Brillas, E., Sires, I. & Oturan, M.A. (2009) Electro-Fenton Process and Related Electrochemical Technologies Based on Fenton's Reaction Chemistry, *Chemical Reviews* 109, American Chemical Society (ACS), 6570-6631.

Canizares, P., Diaz, M., Dominguez, J.A., Garcia-Gomez, J. & Rodrigo, M.A. (2002) Electrochemical Oxidation of Aqueous Phenol Wastes on Synthetic Diamond Thin-film Electrodes, Industrial & Engineering Chemistry Research 41(17), *American Chemical Society* (ACS), 4187-4194.

Canizares, P., Diaz, M., Dominguez, J.A., Lobato, J. & Rodrigo, M.A. (2005) Electrochemical treatment of diluted cyanide aqueous wastes, *Journal of Chemical Technology & Biotechnology* 80(5), Society of Chemical Industry (SCI), Wiley, 565-573.

Canizares, P., Dominguez, J.A., Rodrigo, M.A., Villasenor, J. & Rodriguez, J. (1999) Effect of the Current Intensity in the Electrochemical Oxidation of Aqueous Phenol Wastes at an Activated Carbon and Steel Anode, Industrial & Engineering Chemistry Research 38(10), *American Chemical Society* (ACS), 3779-3785.

Canizares, P., Gadri, A., Lobato, J., Nasr, B., Paz, R., Rodrigo, M.A. & Saez, C. (2006) Electrochemical Oxidation of Azoic Dyes with Conductive-Diamond Anodes, Journal of Industrial & Engineering Chemistry Research 45(10), *American Chemical Society* (ACS), 3468-3473.

Canizares, P., Garcia-Gomez, J., Lobato, J. & Rodrigo, M.A. (2003) Electrochemical Oxidation of Aqueous Carboxylic Acid Wastes Using Diamond Thin-film Electrodes, *Industrial & Engineering Chemistry Research* 42(5), American Chemical Society (ACS), 956-962.

Canizares, P., Garcia-Gomez, J., Saez, C. & Rodrigo, M.A. (2004a) Electrochemical oxidation of several chlorophenols on diamond electrodes: Part II. Influence of waste characteristics and operating conditions, *Journal of Applied Electrochemistry* 34(1), Springer, 87-94.

Canizares, P., Saez, C., Lobato, J. & Rodrigo, M.A. (2004b) Electrochemical treatment of 2,4-dinitrophenol aqueous wastes using boron-doped diamond anodes, *Electrochimica Acta* 49(26), the International Society of Electrochemistry (ISE), 4641-4650.

Cardarelli, F. (2008) Dimensionally Stable Anodes (DSA®) for Chlorine Evolution, Materials Handbook: A Concise Desktop Reference, Springer, 581-582.

Casado, J., Fornaguera, J. & Galan, M.I. (2005) Mineralization of Aromatics in Water by Sunlight-Assisted Electro-Fenton Technology in a Pilot Reactor, *Environmental Science & Technology* 39(6), American Chemical Society (ACS), 1843-1847.

Chandler, G.K., Genders, J.D. & Pletcher, D. (1997) Electrodes Based on Noble Metals, *Platinum Metals Review* 41(2), 54-63.

Chaplin, B.P. (2014) Critical review of electrochemical advanced oxidation processes for water treatment applications, *Environmental Science: Processes & Impacts* 6, Royal Society of Chemistry, 1182-1203.

Chen, G. (2004) Electrochemical technologies in wastewater treatment, *Seperatation and Purification Technology* 38, Elsevier, 11-41.

Chen, G., Chen, X. & Yue, P.L. (2002) Electrochemical behavior of novel Ti/IrO_x-Sb_2O_5-SnO_2 anodes, *Journal of Physical Chemistry* B 106, American Chemical Society (ACS), 4364-4369.

Chen, X., Chen, G. (2005) Investigation of Ti/IrO_2-b_2O_5-SnO_2 electrodes for O_2 evolution calcination temperature and precursor composition effects, *Journal of The Electrochemical Society* 152(7), J59-J64.

Chen, X., Chen, G., Gao, F. & Yue, P.L. (2003) High-performance Ti/BDD electrodes for pollutant oxidation, Environmental Science & Technology 37(21), *American Chemical Society* (ACS), 5021-5026.

Chen, X., Chen, G. & Yue, P.L. (2001) Stable Ti/IrO_x-Sb_2O_5-SnO_2 anode for O_2 evolution with low Ir content, J*ournal of Physical Chemistry* B 105, American Chemical Society (ACS), 4623-4628.

Chen, J., Liu, M., Zhang, J., Xian, Y. & Jin, L. (2003) Electrochemical degradation of bromopyrogallol red in presence of cobalt ions, *Chemosphere* 53(9), 1131-1136.

Chen, F., Yu, S., Dong, X., Zhang, L. & Wu, Q. (2013) Preparation and characterization of PbO_2 electrode and its application in electro-catalytic degradation of o-aminophenol in aqueous solution assisted by CuO-Ce_2O_3/γ-Al_2O_3 catalyst, *Journal of Hazardous Materials* 260, Elsevier, 747-753.

Choi, J., Chung, J. (2014) Reuse of Semiconductor Wastewater Using Reverse Osmosis and Metal-Immobilized Catalyst-Based Advanced Oxidation Process, *Journal of Industrial & Engineering Chemistry Research* 53(27), American Chemical Society (ACS), 11167-11175.

Chu, Y., Zhang, D., Liu, L., Qian, Y. & Li, L. (2013) Electrochemical degradation of m-cresol using porous carbon-nanotube-containing cathode and Ti/SnO_2-Sb_2O_5-IrO_2 anode: Kinetics, byproducts and biodegradability, *Journal of Hazardous Materials* 252-253, Elsevier, 306-312.

Comninellis, C. (1994) Electrocatalysis in the electrochemical conversion/combustion of organic pollutants for wastewater treatment, *Electrochimica Acta* 39(11-12), The International Society of Electrochemistry (ISE), Elsevier, 1857-1862.

Comninellis, C., Chen, G. (2010) *Electrochemistry for the Environment,* Springer, New York-Dordrecht-Heidelberg-London.

Comninellis, C., Duo, I., Michaud, P.A., Marselli, B. & Park, S.M. (2005) Application of synthetic boron-doped diamond electrodes in electrooxidation processes, In: *Diamond Electrochemistry, Fujishima et al.* (Eds.), Elsevier, Amsterdam, 449-476.

Comninellis, C., Kapalka, A., Malato, S., Parsons, S.A., Poulios, I. & Mantzavinos, D. (2008) Advanced oxidation processes for water treatment: advances and trends for R&D, *Journal of Chemical Technology and Biotechnology* 83(6), Society of Chemical Industry (SCI), 769-776.

Comninellis, C., Nerini, A. (1995) Anodic oxidation of phenol in the presence of NaCl for wastewater treatment, *Journal of Applied Electrochemistry* 25, Springer, 23-28.

Comninellis, C., Pulgarin, C. (1991) Anodic oxidation of phenol for wastewater treatment, *Journal of Applied Electrochemistry* 21, Springer, 703-708.

Comninellis, C., Pulgarin, C. (1993) Electrochemical oxidation of phenol for wastewater treatment using SnO_2 anodes, *Journal of Applied Chemistry* 23(2), Springer, 108-112.

Correa-Lozano, B., Comninellis, C. & De Battisti, A. (1997) Service life of Ti/SnO_2-Sb_2O_5 anodes, *Journal of Applied Electrochemistry* 27(8), Springer, 970-974.

Cossu, R., Polcaro, A.M., Lavagnolo, M.C., Mascia, M., Palmas, S. & Renoldi, F. (1998) Electrochemical treatment of landfill leachate : Oxidation at Ti/PbO_2 and Ti/SnO_2 anodes, Journal of Industrial & Engineering Chemistry Research 32, *American Chemical Society* (ACS), 3570-3573.

Costa, C.R., Botta, C.M.R., Espindola, E.L.G. & Olivi, P. (2008) Electrochemical treatment of tannery wastewater using DSA® electrodes, *Journal of Hazardous Materials* 153(1-2), Elsevier, 616-627.

Costa, C.R., Montilla, F., Morallon, E. & Olivi, P. (2010) Electrochemical oxidation of synthetic tannery wastewater in chloride-free aqueous media, *Journal of Hazardous Materials* 180(1-3), Elsevier, 429-435.

Cui, Y.H., Feng, Y.J., Liu, J. & Ren, N. (2012) Comparison of various organic compounds destruction on rare earths doped Ti/Sb-SnO_2 electrodes, *Journal of Hazardous Materials* 239-240, Elsevier, 225-232.

Cui, Y.H., Li, X.Y. & Chen, G. (2009) Electrochemical degradation of bisphenol A on different anodes, *Water Research* 43(7), the International Water Association (IWA), Elsevier, 1968-1976.

Dai, Q., Wang, J., Chen, L. & Chen, J. (2014) Degradation of p-Acetamidophenol in Aqueous Solution by Ozonation: Performance Optimization and Kinetics Study, Journal of Industrial & Engineering Chemistry Research 53(29), *American Chemical Society* (ACS), 11593-11600.

Da Pozzo, A., Di Palma, L., Merli, C. & Petrucci, E. (2005) An experimental comparison of a graphite electrode and a gas diffusion electrode for the cathodic production of hydrogen peroxide, *Journal of Applied Electrochemistry* 35(4), Springer, 413-419.

Da Pozzo, A., Petrucci, E. & Merli, C. (2008) Electrogeneration of hydrogen peroxide in seawater and application to disinfection, *Journal of Applied Electrochemistry* 38(7), Springer, 997-1003.

De Leon, C.P., Pletcher, D. (1995) Removal of formaldehyde from aqueous solutions via oxygen reduction using a reticulated vitreous carbon cathode cell, *Journal of Applied Electrochemistry* 25(4), Springer, 307-314.

De Luis, A.M., Lombrana, J.I., Menendez, A. & Sanz, J. (2011) Analysis of the Toxity of Phenol Solutions Treated with H_2O_2/UV and H_2O_2/Fe Oxidative Systems, Journal of Industrial & Engineering Chemistry Research 50(4), *American Chemical Society* (ACS), 1928-1937.

Debiemme-Chouvy, C., Hua, Y., Hui, F., Duval, J.L. & Cachet, H. (2011) Electrochemical treatments using tin oxide anode to prevent biofouling, *Electrochimica Acta* 56(28), the International Society of Electrochemistry (ISE), Elsevier, 10364-10370.

Do, J.S., Chen, C.P. (1993) In situ oxidative degradation of formaldehyde with electrogenerated hydrogen peroxide, *Journal of the Electrochemical Society,* 140(6), 1632-1637.

Do, J.S., Chen, C.P. (1994) Kinetics of in situ degradation of formaldehyde with electrogenerated hydrogen peroxide, *Journal of Industrial & Engineering Chemistry Research* 33(2), American Chemical Society (ACS), 387-394.

Do, J.S., Chen, C.P. (1994) In situ oxidative degradation of formaldehyde with hydrogen peroxide electrogenerated on the modified graphites, *Journal of Applied Electrochemistry* 24(9), Springer, 936-942.

Do, J.S., Yeh, W.C. (1995) In situ degradation of formaldehyde with electrogenerated hypochlorite ion, *Journal of Applied Electrochemistry* 25(5), Springer, 483-489.

Do, J.S., Yeh, W.C. (1996) Paired electrooxidative degradation of phenol with in situ electrogenerated hydrogen peroxide and hypochlorite, *Journal of Applied Electrochemistry* 26(6), Springer, 673-678.

Do, J.S., Yeh, W.C. (1998) In Situ paired electrooxidative degradation of formaldehyde with electrogenerated hydrogen peroxide and hypochlorite ion, *Journal of Applied Electrochemistry* 28(7), Springer, 703-710.

Do, J.S., Yeh, W.C. & Chao, I.Y. (1997) Kinetic of the oxidative degradation of formaldehyde with electrogenerated hypochlorite ion, Journal of Industrial & Engineering Chemistry Research 36(2), *American Chemical Society* (ACS), 349-356.

Dong, X., Wang, Y., Li, X., Yu, Y. & Zhang, M. (2014) Process Simulation of Laboratory Wastewater Treatment via Supercritical Water Oxidation, Journal of Industrial & Engineering Chemistry Research 53(18), *American Chemical Society* (ACS), 7723-7729.

Duan, T., Wen, Q., Chen, Y., Zhou, Y. & Duan, Y. (2014) Enhancing electrolytic performance of Sb-doped SnO_2 electrode by compositing nitrogen-doped graphene nanosheets, *Journal of Hazardous Materials* 280, Elsevier, 304-314.

El-Ghenymy, A., Garrido, J.A., Centellas, F., Arias, C., Cabot, P.L., Rodriguez, R.M. & Brillas, E. (2012) Electro-Fenton and Photoelectro-Fenton Degradation of Sulfanilic Acid Using a Boron-Doped Diamond Anode and an Air Diffusion Cathode, T*he Journal of Physical Chemistry A* 16(13), American Chemical Society (ACS), 3404-3412.

Esquivel, K., Arriaga, L.G., Rodriguez, F.J., Martinez, L. & Godinez, L.A. (2009) Development of a TiO_2 modified optical fiber electrode and its incorporation into a photoelectrochemical reactor for wastewater treatment, *Water Research* 43(14), the International Water Association (IWA), Elsevier, 3593-3603.

Fan, Y., Ai, Z. & Zhang, L. (2010) Design of an electro-Fenton system with a novel sandwich film cathode for wastewater treatment, *Journal of Hazardous Materials* 176(1-3), Elsevier, 678-684.

Farhadi, S., Aminzadeh, B., Torabian, A., Khatibikamal, V. & Alizadeh Fard, M. (2012) Comparison of COD removal from pharmaceutical wastewater by electrocoagulation, photoelectrocoagulation, peroxi-electrocoagulation and peroxi-photoelectrocoagulation processes, *Journal of Hazardous Materials* 219-220, Elsevier, 35-42.

Feng, C.H., Li, F.B., Mai, H.J. & Li, X.Z. (2010) Bio-Electro-Fenton Process Driven by Microbial Fuel Cell for Wastewater Treatment, Environmental Science & Technology 44(5), *American Chemical Society* (ACS), 1875-1880.

Feng, L., Oturan, N., van Hullebusch, E.D., Esposito, G. & Oturan, M.A. (2014) Degradation of anti-inflammatory drug Ketoprofen by electro-oxidation: comparison of electro-Fenton and anodic oxidation process, *Journal of Environmental Sience and Pollution Research* 21(14), Springer, 8406-8416.

Fernandes, A., Morao, A., Magrinho, M., Lopes, A. & Goncalves, I. (2004) Electrochemical degradation of C.I. Acid Orange 7, *Dyes and Pigments* 61(3), Elsevier, 287-296.

Ferrag-Siagh, F., Fourcade, F., Soutrel, I., Ait-Amar, H., Djelal, H. & Amrane, A. (2014) Electro-Fenton pretreatment for the improvement of tylosin biodegradability, *Journal of Environmental Science and Pollution Research* 21(14), Springer, 8534-8542.

Flox, C., Cabot, P.L., Centellas, F., Garrido, J.A., Rodriguez, R.M., Arias, C. & Brillas, E. (2007) Solar photoelectron-Fenton degradation of cresols using a flow reactor with a boron-doped diamond anode, *Applied Catalysis B: Environmental* 75(1-2), Elsevier, 17-28.

Foti, G., Gandini, D. & Comninellis, C. (1997) Anodic oxidation of organics on thermally prepared oxide electrodes, *Current Topics in Electrochemistry* 5, Research Trends, 71-91.

Foti, G., Gandini, D., Comninellis, C., Perret, A. & Haenni, W. (1999) Oxidation of organics by intermediates of water discharge on IrO_2 and synthetic diamond anodes, *Electrochemical and Solid-State Letters* 2(5), the society for solid-state and electrochemical science and technology (ECS), 228-230.

Fu, J., Zhang, X., & Lei, L. (2007) Fe-modified multi-walled carbon nanotube electrode for production of hydrogen peroxide, *Acta Physico-Chimica Sinica* 23(8), 1157-1162.

Gallard, H., De Laat, J., & Legube, B. (1999) Etude comparative de la vitesse de décomposition de H_2O_2 et de l'atrazine par les systèmes Fe (III)/H_2O_2, Cu (II)/H_2O_2 et Fe (III)/Cu (II)/H_2O_2, *Revue des sciences de l'eau/Journal of Water Science,* 12(4), 713-728.

Gallegos, A.A., García, Y.V. & Zamudio, A. (2005) Solar hydrogen peroxide, *Solar energy materials and solar cells,* 88(2), 157-167.

Gandini, D., Comninellis, C., Perret, A. & Haenni, W. (1999) Anodic oxidation of organics on synthetic diamond thin-film electrodes, *ICHEME* Symposium Series 145, 181-190.

Gandini, D., Mahe, E., Michaud, P.A., Haenni, W., Perret, A. & Comninellis, C. (2000) Oxidation of carboxylic acids at boron-doped diamond electrodes for wastewater treatment, *Journal of Applied Electrochemistry* 30(12), Springer, 1345-1350.

Ganzenko, O., Huguenot, D., van Hullebusch, E.D., Esposito, G. & Oturan, M.A. (2014) Electrochemical advanced oxidation and biological processes for wastewater treatment: a review of the combined approaches, Journal *of Environmental Science and Pollution Research* 21(14), Springer, 8493-8524.

Garcia-Segura, S., Garrido, J.A., Rodriguez, R.M., Cabot, P.L., Centellas, F., Arias, C. & Brillas, E. (2012) Mineralization of flumequine in acidic medium by electro-Fenton and photoelectron-Fenton processes, *Water Research* 46(7), The International Water Association (IWA), Elsevier, 2067-2076.

Gattrel, M., Kirk, D. (1993) A Study of the Oxidation of phenol at Platinum and Preoxidized Platinum Surfaces, *Journal of The Electrochemical Society* 140(6), the society for solid-state and electrochemical science and technology (ECS), 1534-1540.

Gherardini, L., Michaud, P.A., Panizza, M., Comninellis, C. & Vatistas, N. (2001) Electrochemical oxidation of 4-chlorophenol for wastewater treatment. Definition of normalized current efficiency, *Journal of The Electrochemical Society* (JES) 148(6), The society for solid-state and electrochemical science and technology (ECS), D78-D82.

Goncalves, M.R., Marques, I.P. & Correia, J.P. (2012) Electrochemical mineralization of anaerobically digested olive mill wastewater, Water research 46(13), *International Water Association* (IWA), Elsevier, 4217-4225.

Grimm, J.H., Bessarabov, D.G., Simon, U. & Sanderson, R.D. (2000) Characterization of doped tin dioxide anodes prepared by a sol-gel technique and their application in an SPE-reactor, *Journal of Applied Electrochemistry* 30(3), Springer, 293-302.

Gunagi, S.D., Naik, P.N., Nandibewoor, S.T. & Chimatadar, S.A. (2010) Oxidation of 6-Aminopenicillanic Acid by Diperiodatoagantate (III) in Aqueous Alkaline Medium: A Kinetic and Mechanistic Study, *Journal of Industrial & Engineering Chemistry Research* 49(20), American Chemical Society (ACS), 9666-9672.

Guo, H., Feng, S., Jiang, J., Zhang, M., Lin, H. & Zhou, X. (2014) Application of Fenton's reagent combined with sawdust on the dewaterability of oily sludge, *Journal of Environmental Science and Pollution Research* 21(18), Springer, 10706-10712.

Guo, L., Li, X. & Chen, G. (2010) Techniques of Electrode Fabrication, *Electrochemistry for the Environment,* Springer, 55-98.

Hamidi, A.A., Othman, O.M. & Abu Amr, S.S. (2013) The performance of Electro-Fenton oxidation in the removal of coliform bacteria from landfill leachate, *Waste Management* 33(2), Elsevier, 396-400.

Hamni, E., Elaoud, S.C., Samet, Y. & Abdelhedi, R. (2009) Electrochemical degradation of waters containing O-Toluidine on PbO_2 and BDD anodes, *Journal of Hazardous Materials* 170(2-3), Elsevier, 928-933.

Harrington, T., Pletcher, D. (1999) The Removal of Low Levels of Organics from Aqueous Solutions Using Fe (II) and Hydrogen Peroxide Formed In Situ at Gas Diffusion Electrodes, *Journal of The Electrochemical Society* (JES) 146(8), the society for solid-state and electrochemical science and technology, 2983-2989.

Hattori, S., Doi, M., Takahashi, E., Kurosu, T., Nara, M., Nakamatsu, S., Nishiki, Y., Furuta, T. & Iida, M. (2003) Electrolytic decomposition of amaranth dyestuff using diamond electrodes, *Journal of Applied Electrochemistry* 33(1), Springer, 85-91.

He, Z., Cai, Q., Hong, F., Jiang, Z., Chen, J. & Song, S. (2012) Effective Enhancement of the Degradation of Oxalic Acid by Catalytic Ozonation with TiO_2 by Exposure of {001} Facets and Surface Fluorination, J*ournal of Industrial & Engineering Chemistry Research* 51(16), American Chemical Society (ACS), 5662-5668.

He, Z., Gao, C., Qian, M., Shi, Y., Chen, J. & Song, S. (2014) Electro-Fenton Process Catalyzes by Fe_3O_4 Magnetic Nanoparticles for Degradation of C.I. Reactive Blue 19 in Aqueous Solution: Operating Conditions, Influence, and Mechanism, *Journal of Industrial & Engineering Chemistry Research* 53(9), American Chemical Society (ACS), 3435-3447.

Hernandez, M.C., Russo, N., Panizza, M., Spinelli, D. & Fino, D. (2014) Electrochemical oxidation of urea in aqueous solutions using a boron-doped thin-film diamond electrode, *Diamond and Related Materials* 44, Elsevier, 109-116.

Ho, C.C., Chan, C.Y. & Khoo, K.H. (2007) Electrochemical treatment of effluents: A preliminary study of anodic oxidation of simple sugars using lead dioxide-coated titanium anodes, Journal of Chemical Technology and Biotechnology 36(1), *Society of Chemical Industry (*SCI), Wiley, 7-14.

Houk, L.L., Johnson, S.K., Feng, J., Houk, R.S. & Johnson, D.C. (1998) Electrochemical incineration of benzoquinone in aqueous media using a quaternary metal oxide electrode in the absence of a soluble supporting electrolyte, *Journal of Applied Electrochemistry* 28(11), Springer, 1167-1177.

Hu, Z., Zhou, M., Zhou, L., Li, Y. & Zhang, C. (2014) Effect of matrix on the electrochemical characteristics of TiO_2 nanotube array based PbO_2 electrode for pollution degradation, *Journal of Environmental Science and Pollution Research* 21(14), Springer, 8476-8484.

Huang, C.P., Dong, C. & Tang, Z. (1993) Advanced chemical oxidation: its present role and potential future in hazardous waste treatment, *Waste Management* 13, 361-377.

Huang, W., Ji, Y., Yang, Z., Feng, X., Liu, C., Yinhua, Z. & Lu, X. (2010) Mineralization of Trace Nitro/Chloro/Methyl/Amino-Aromatic Contaminants in Wastewaters by Advanced

Oxidation Processes, *Journal of Industrial & Engineering Chemistry Research* 49(13), American Chemical Society (ACS), 6243-6249.

Iglesias, I., Fernandez de Dios, M.A., Rosales, E., Pazos, M. & Sanroman, M.A. (2013) Optimization of decolourisation and degradation of Reactive Black 5 dye under electro-Fenton process using Fe alginate gel beads, *Journal of Environmental Science and Pollution Research* 20(4), Springer, 2172-2183.

Iglesias, O., Sanroman, M.A. & Pazos, M. (2014) Surfactant-Enhanced Solubilization and Simultaneous Degradation of Phenanthrene in Marine Sediment by Electro-Fenton Treatment, J*ournal of Industrial & Engineering Chemistry Research* 53(8), American Chemical Society (ACS), 2917-2923.

Iniesta, J., Michaud, P.A., Panizza, M. & Comninellis, C. (2001) Electrochemical oxidation of 3-methylpyridine at a boron-doped diamond electrode : Application to electroorganic synthesis and wastewater treatment, *Electrochemistry Communications* 3(7), Elsevier, 346-351.

Johnson, S.K., Houk, L.L., Feng, J., Houk, R.S. & Johnson, D.C. (1999) Electrochemical incineration of 4-chlorophenol and the identification of products and intermediates by mass spectrometry, *Environmental Science & Technology* 33, American Chemical Society (ACS), 2638-2644.

Kang, W.J., Glaze, W.H. & Chapin, D.H. (1987) The Chemistry of Water Treatment Processes Involving Ozone, Hydrogen Peroxide and Ultraviolet Radiation, *Ozone Science and Engineering* 9(4), International Ozone Association, 335-352.

Karuppiah, J., Reddy, E.L., Reddy, P.M.K., Ramaraju, B., Karvembu, R. & Subrahmanyam, Ch. (2012) Abatement of mixture of volatile organic compounds (VOCs) in a catalytic non-thermal plasma reactor, *Journal of Hazardous Materials* 237-238, Elsevier, 283-289.

Keech, P.G., Bunce, N.J. (2003) Electrochemical oxidation of simple indoles at a PbO_2 anode, *Journal of Applied Electrochemistry* 33(1), Springer, 79-83.

Khoufi, S., Aloui, F. & Sayadi, S. (2006) Treatment of olive oil mill wastewater by combined process electro-Fenton reaction and anaerobic digestion, *Water Research* 40(10), International Water Association (IWA), Elsevier, 2007-2016.

Kishimoto, N., Kitamura, T., Masaaki, K. & Otsu, H. (2013) Reusability of iron sludge as an iron source for the electrochemical Fenton-type process using Fe^{2+}/HOCl system, *Water Research* 47(5), International Water Association (IWA), Elsevier, 1919-1927.

Koetz, R., Stucki, S. & Carcer, B. (1991) Electrochemical wastewater treatment using high voltage anodes. Part I: Physical and electrochemical properties of SnO_2 anodes, *Journal of Applied Electrochemistry* 21(1), Springer, 14-20.

Koivula, R. (2010) The effect of dopant's valence (+III and +V)on the anion/cation uptake properties of antimony-doped tin dioxide, *Central European Journal of Chemistry* 8(6), Springer, 1179-1184.

Kornienko, G.V., Chaenko, N.V. & Vasil'eva, I.S. (2004) Indirect Electrooxidation of organic Substrates by Hydrogen Peroxide Generated in an Oxygen-diffusion Electrode, *Russian Journal of Electrochemistry* 41(7), Springer, 148-152.

Kraft, A., Stadelmann, M. & Blaschke, M. (2003) Anodic oxidation with doped diamond electrodes: A new advanced oxidation process, *Journal of Hazardous Materials 1*03(3), Elsevier, 247-261.

Krzywicka, A., Kwarciak-Kozlowski, A. (2014) Advanced oxidation processes with coke plant wastewater treatment, *Water Science & Technology, I*nternational Water Association (IWA).

Laiju, A.R., Sivasankar, T. & Nidheesh, P.V. (2014) Iron-loaded mangosteen as a heterogeneous Fenton catalyst for the treatment of landfill leachate, *Journal of Environmental Science and Pollution Research* 21(18), Springer, 10900-10907.

Lamy, C. (1984) Electrocatalytic oxidation of organic compounds on noble metals in aqueous solution, *Electrochimica Acta* 29(11), the International Society of Electrochemistry (ISE), Elsevier, 1581-1588.

Lamy, C., Leger, J.M., Clavilier, J. & Parsons, R. (1983) Structural effects in electrocatalysis: A comparative study of the oxidation of CO, HCOOH and CH_3OH on single crystal Pt electrodes, *Journal of Electroanalytical Chemistry* 150(1-2), the International Society of Electrochemistry (ISE), Elsevier, 71-77.

Lanza, M.R.V., Bertazzoli, R. (2002) Cyanide Oxidation from Wastewater in a Flow Electrochemical Reactor, Journal of Industrial & Engineering Chemistry Research 41(1), *American Chemical Society (*ACS), 22-26.

Lee, H., Yoo, H.Y., Choi, J., Nam, I.H., Lee, S., Lee, S., Kim, J.H., Lee, C. & Lee, J. (2014) Oxidizing Capacity of Periodate Activated with Iron-Based Bimetallic Nanoparticles, *Environmental Science & Technology* 48(14), American Chemical Society (ACS), 8086-8093.

Lei, L., Dai, Q. (2007) High Performance on the Degradation of Cationic Red X-GRL by Wet Electrocatalytic Oxidation Process, *Journal of Industrial & Engineering Chemistry Research* 46(26), American Chemical Society (ACS), 8951-8958.

Lei, Y., Liu, H., Shen, Z. & Wang, W. (2013) Development of a trickle bed reactor of electro-Fenton process for wastewater treatment, Journal of *Hazardous Materials* 261, Elsevier, 570-576.

Li, G., Wang, Y.H. & Chen, Q.Y. (2013) Influence of fluoride-doped tin oxide interlayer on Ni-Sb-SnO_2/Ti electrodes, *Journal of Solid State Electrochemistry* 17(5), Springer, 1303-1309.

Li, H., Gong, Y., Huang, Q. & Zhang, H. (2013) Degradation of Orange II by UV-assisted Advanced Fenton Process: Response Surface Approach, Degradation Pathway, and Biodegradability, *Journal of Industrial & Engineering Chemistry Research 52*(44), American Chemical Society (ACS), 15560-15567.

Li, J., Ai, Z. & Zhang, L. (2009) Design of a neutral electro-Fenton system with Fe@Fe_2O_3/ACF composite cathode for wastewater treatment, *Journal of Hazardous Materials* 164(1), Elsevier, 18-25.

Li, X.Y., Cui, Y.H., Feng, Y.J., Xie, Z.M. & Gu, J.D. (2005) Reaction pathways and mechanisms of the electrochemical degradation of phenol on different electrodes, *Water Research* 39(10), the International Water Association (IWA), Elsevier, 1972-1981.

Liao, P., Yuan, S., Chen, M., Tong, M., Xie, W. & Zhang, P. (2013) Regulation of Electrochemically Generated Ferrous Ions from an Iron Cathode for Pd-Catalytic Transformation of MTBE in Groundwater, Environmental Science & Technology 47(14), *American Chemical Society* (ACS), 7918-7926.

Lipp, L., Pletcher, D. (1997) Preparation and Characterization of tin dioxide coated titanium electrodes, *Electrochimica Acta* 42(7), the International Society of Electrochemistry (ISE), Elsevier, 1091-1099.

Lissens, G., Pieters, J., Verhaege, M., Pinoy, L. & Verstraete, W. (2003) Electrochemical degradation of surfactants by intermediates of water discharge at carbon-based electrodes, *Electrochimica Acta* 48(12), the International Society of Electrochemistry (ISE), Elsevier, 1655-1663.

Liu, W., Ai, Z. & Zhang, L. (2012) Design of a neutral three-dimensional electro-Fenton system with foam nickel as particle electrodes for wastewater treatment, *Journal of Hazardous Materials* 243, Elsevier, 257-264.

Liu, H., Li, X.Z., Leng, Y.J., & Wang, C. (2007) Kinetic modeling of electro-Fenton reaction in aqueous solution, *Water research* 41(5), the International Water Association (IWA), Elsevier, 1161-1167.

Liu, Y., Liu, H. & Li, Y. (2008) Comparative study of the electrocatalytic oxidation and mechanism of nitrophenols at Bi-doped lead dioxide anodes, *Journal of Applied Catalysis B: Environmental* 84(1), Elsevier, 297-302.

Liu, Y., Liu, H., Ma, J. & Li, J. (2011) Investigation on electrochemical properties of cerium doped lead dioxide anode and application for elimination of nitrophenol, *Electrochimica Acta* 56(3), the International Society of Electrochemistry (ISE), Elsevier, 1352-1360.

Liu, Y., Liu, H., Ma, J. & Li, J. (2012) Preparation and electrochemical properties of Ce-Ru-SnO_2 ternary oxide anode and electrochemical oxidation of nitrophenols, *Journal of Hazardous Materials* 213-214, Elsevier, 222-229.

Magureanu, M., Piroi, D., Mandache, N.B., David, V., Medvedovici, A. & Parvulescu, V.I. (2010) Degradation of pharmaceutical compound pentoxifylline in water by non-thermal plasma treatment, *Water Research* 44(11), the International Water Association (IWA), Elsevier, 3445-3453.

Magureanu, M., Piroi, D., Mandache, N.B., David, V., Medvedovici, A., Bradu, C. & Parvulescu, V.I. (2011) Degradation of antibiotics in water by non-thermal plasma treatment, *Water Research* 45(11), the International Water Association (IWA), Elsevier, 3407-3416.

Malpass, G.R.P., Neves, R.S. & Motheo, A.J. (2006) A comparative study of commercial and laboratory-made $Ti/Ru_{0.3}Ti_{0.7}O_2$ DSA electrodes. "In situ" and "ex situ" surface characterization and organic oxidation activity, *Electrochimica Acta* 52, the International Society of Electrochemistry (ISE), Elsevier, 936-944.

Martinez, S.S., Bahena, C.L. (2009) Chlorbromuron urea herbicide removal by electro-Fenton reaction in aqueous effluents, *Water Research* 43(1), the International Water Association (IWA), Elsevier, 33-40.

Martinez-Huitle, C.A. (2014) Organic Pollutants in Water Using BDD, Direct and Indirect Electrochemical Oxidation, *Encyclopedia of Applied Electrochemistry,* Springer, 1402-1407.

Martinez-Huitle, C.A., Quiroz, M.A., Comninellis, C., Ferro, S. & De Battisti, A. (2004) Electrochemical incineration of chloranilic acid using Ti/IrO_2, Pb/PbO_2 and Si/BDD electrodes, *Electrochimica Acta* 50(4), the International Society of Electrochemistry (ISE), Elsevier, 949-956.

Milovac, N., Juretic, D., Kusic, H., Dermadi, J. & Bozic, A.L. (2014) Photooxidative Degradation of Aromatic Carboxylic Acids in Water: Influence of Hydroxyl Substituents, *Journal of Industrial & Engineering Chemistry Research* 53(26), American Chemical Society (ACS), 10590-10598.

Molkenthin, M., Olmez-Hanci, T., Jekel, M.R. & Arslan-Alaton, I. (2013) Photo-Fenton-Like treatment of BPA: Effect of UV light source and water matrix on toxicity and transformation products, *Water Research* 47(14), the International Water Association (IWA), Elsevier, 5052-5064.

Montilla, F., Michaud, P.A., Morallon, E., Vazquez, J.L. & Comninellis, C. (2002) Electrochemical oxidation of benzoic acid at boron-doped diamond electrodes, *Electrochimica Acta* 47(21), the International Society of Electrochemistry (ISE), Elsevier, 3509-3513.

Montilla, F., Morallon, E., De Battisti, A. & Vazquez, J.L. (2004a) Preparation and Characterization of Antimony-Doped Tin Oxide Electrodes. Part 1. Electrochemical Characterization, *The Journal of Physical Chemistry B* 108(16), American Chemical Society (ACS), 5036-5043.

Montilla, F., Morallon, E., De Battisti, A., Benedetti, A., Yamashita, H. & Vazquez, J.L. (2004b) Preparation and Characterization of Antimony-Doped Tin Dioxide Electrodes. Part 2. XRD and EXAFS Characterization, *The Journal of Physical Chemistry* B 108(16), American Chemical Society (ACS), 5044-5050.

Montilla, F., Morallon, E., De Battisti, A., Barison, S., Daolio, S. & Vazquez, J.L. (2004c) Preparation and Characterization of Antimony-Doped Tin Dioxide Electrodes, Part 3, XPS and SIMS Characterization, *The Journal of Physical Chemistry* B 108(41), American Chemical Society (ACS), 15976-15981.

Montilla, F., Morallon, E. & Vazquez, J.L. (2005) Evaluation of the Electrocatalytic Activity of Antimony-Doped Tin Dioxide Anodes toward the oxidation of Phenol in Aqueous Solutions, *Journal of The Electrochemical Society* (JES) 152(10), the society for solid-state and electrochemical science and technology (ECS), B421-B427.

Morao, A., Lopes, A., Amorim, M.T.P. & Goncalves, I.C. (2004) Degradation of mixtures of phenols using boron-doped diamond electrodes for wastewater treatment, *Electrochimica Acta* 49(9-10), the International Society of Electrochemistry (ISE), Elsevier, 1587-1595.

Mozgina, O., Koutsospyros, A., Gershman, S., Belkind, A., Christodoulatos, C. & Becker, K.H. (2009) Decomposition of Energetic Materials by Pulsed Electrical Discharges in Gas-Bubbled Aqueous Solutions, *IEEE Transactions on Plasma Science* 37(6), 905-910.

Muhammad, S., Shukla, P.R., Tade, M.O. & Wang, S. (2012) Heterogeneous activation of peroxymonosulphate by supported ruthenium catalysts for phenol degradation in water, *Journal of Hazardous Materials* 215-216, Elsevier, 183-190.

Muhammad, U., Hamidi, A.A. & Mohd. Suffian, Y. (2010) Trends in the use of Fenton, electro-Fenton and photo-Fenton for the treatment of landfill leachate, *Waste Management* 30, Elsevier, 2113-2121.

Nidheesh, P.V., Gandhimathi, R. (2012) Trends in electro-Fenton process for water and wastewater treatment: An overview, *Desalination* 299, Elsevier, 1-15.

Nidheesh, P.V., Gandhimathi, R. (2014) Electrolytic removal of Rhodamine B from aqueous solution by peroxicoagulation process, *Environmental Science and Pollution Research* 21(14), Springer, 8585-8594.

Ochiai, T., Lizuka, Y., Nakata, K., Murakami, T., Tryk, D.A., Fujishima, A., Koide, Y. & Morito, Y. (2011) Efficient electrochemical decomposition of perfluorocarboxylic acids by the use of a boron-doped diamond electrode, *Diamond and Related Materials* 20(2), Elsevier, 64-67.

Oeguetveren, U.B., Toru, E. & Koparal, S. (1999) Removal of cyanide by anodic oxidation for wastewater treatment, *Water Research* 33(8), the International Water Association (IWA), Elsevier, 1851-1856.

Oezcan, A., Sahin, Y., Koparal, A.S., & Oturan, M.A. (2008) Carbon sponge as a new cathode material for the electro-Fenton process: comparison with carbon felt cathode and application to degradation of synthetic dye basic blue 3 in aqueous medium, *Journal of Electroanalytical Chemistry* 616(1), 71-78.

Oturan, M. (1999) Hydroxylation of aromatic drugs by the electro-Fenton method. Formation and identification of the metabolites of Riluzole, *New Journal of Chemistry* 23(8), 793-794.

Oturan, M.A. (2000) An ecologically effective water treatment technique using electrochemically generated hydroxyl radicals for in situ destruction of organic pollutants: Application to herbicide 2, 4-D, *Journal of Applied Electrochemistry* 30(4), 475-482.

Oturan, M.A., Aaron, J.J., Oturan, N., & Pinson, J. (1999) Degradation of chlorophenoxyacid herbicides in aqueous media, using a novel electrochemical method, *Pesticide science* 55(5), Pesticide Science Society of Japan, 558-562.

Oturan, N., Panizza, M. & Oturan, M.A. (2009) Cold Incineration of Chlorophenols in Aqueous Solution by Advanced Electrochemical Process Electro-Fenton. Effect of Number and Position of Chlorine Atoms on The Degradation Kinetics, *The Journal of Physical Chemistry* A 113(41), American Chemical Society (ACS), 10988-10993.

Oturan, N., Zhou, M. & Oturan, M.A. (2010) Metomyl Degradation by Electro-Fenton and Electro-Fenton-Like Processes: A Kinetics Study of Effect of the Nature and Concentration of Some Transition Metal Ions as Catalyst, *The Journal of Physical Chemistry* A 114(39), American Chemical Society (ACS), 10605-10611.

Panizza, M. (2010) Importance of Electrode Material in the Electrochemical Treatment of Wastewater Containing Organic Pollutants, In: *Electrochemistry for the Environment,* Springer, 25-54.

Panizza, M., Brillas, E. & Comninellis, C. (2008) Application of Boron-Doped Diamond Electrodes for Wastewater Treatment, *Journal of Environmental Engineering and Management* 18(3), 139-153.

Panizza, M., Cerisola, G. (2003) Electrochemical oxidation of 2-naphtol with in situ electrogenerated active chlorine, *Electrochimica Acta* 48(11), the International Society of Electrochemistry (ISE), Elsevier, 1515-1519.

Panizza, M., Cerisola, G. (2005) Application of diamond electrodes to electrochemical processes, *Electrochimica Acta* 51(2),the International Society of Electrochemistry (ISE), Elsevier, 191-199.

Panizza, M., Cerisola, G. (2006) Electrochemical oxidation of aromatic sulphonated acids on a boron-doped diamond electrode, *International Journal of Environment and Pollution* (IJEP) 27(1/2/3), Inderscience Publishers, 64-74.

Panizza, M., Cerisola, G. (2008) Electrochemical generation of H_2O_2 in low ionic strength media on gas diffusion cathode fed with air, Electrochimica Acta 54(2), *the International Society of Electrochemistry* (ISE), Elsevier, 876-878.

Panizza, M., Cerisola, G. (2010) Applicability of electrochemical methods to carwash wastewaters for reuse, Part 1: Anodic oxidation with diamond and lead dioxide anodes,

Journal of Electroanalytical Chemistry 638(1), International Society of Electrochemistry (ISE), Elsevier, 28-32.

Panizza, M., Michaud, P.A., Cerisola, G. & Comninellis C. (2001a) Anodic oxidation of 2-naphthol at boron-doped diamond electrodes, *Journal of Electroanalytical Chemistry* 507(1-2), the International Society of Electrochemistry (ISE), Elsevier, 206-214.

Panizza, M., Michaud, P.A., Cerisola, G. & Comninellis C. (2001b) Electrochemical treatment of wastewater containing organic pollutants on boron-doped diamond electrodes. Prediction of specific energy consumption and required electrode area, *Electrochemistry Communications* 3(7), the International Society of Electrochemistry, Elsevier, 336-339.

Perret, A., Haenni, W., Skinner, N., Tang, X.M., Gandini, D., Comninellis, C., Correa, B. & Foti, G. (1999) Electrochemical behavior of synthetic diamond thin film electrodes, *Diamond and Related Materials* 8(2-5), Elsevier, 820-823.

Pimentel, M., Oturan, N., Dezotti, M., & Oturan, M. A. (2008) Phenol degradation by advanced electrochemical oxidation process electro-Fenton using a carbon felt cathode, *Applied Catalysis B: Environmental* 83(1), 140-149.

Pignatello, J.J., Oliveros, E. & MacKay, A. (2006) Advanced Oxidation Processes for Organic Contaminant Destruction Based on the Fenton Reaction and Related Chemistry, *Critical Reviews in Environmental Science and Technology* 36(1), Taylor & Francis, 1-84.

Pipi, A.R.F., Neto, S.A. & De Andrade, A.R. (2013) Electrochemical degradation of diuron in chloride medium using DSA® based anodes, *Journal of the Brasilian Chemical Society* 24(8), 1259-1266.

Piya-areetham, P., Shenchunthichai, K. & Hunsom, M. (2006) Application of electrooxidation process for treating concentrated wastewater from distillery industry with a voluminous electrode, Water *Research* 40(15), the International Water Association (IWA), Elsevier, 2857-2864.

Plakas, K.V., Karabelas, A.J., Sklari, S.D. & Zaspalis, V.T. (2013) Toward the Development of a Novel Electro-Fenton System for Eliminating Toxic Organic Substances from Water. Part 1. In Situ Generation of Hydrogen Peroxide, *Journal of Industrial & Engineering Chemistry Research* 52(39), American Chemical Society (ACS), 13948-13956.

Pletcher, D. & Walsh, F.C. (1982) *Industrial Electrochemistry,* Chapman and Hall, London.

Polcaro, A.M., Palmas, S., Renoldi, F. & Mascia, M. (1999) On the performance of Ti/SnO_2 and Ti/PbO_2 anodes in electrochemical degradation of 2-chlorophenol for wastewater treatment, *Journal of Applied Electrochemistry* 29(2), Springer, 147-151.

Polcaro, A.M., Palmas, S. (1997) Electrochemical oxidation of chlorophenols, *Journal of Industrial & Engineering Chemistry Research* 36(5), American Chemical Society (ACS), 1791-1798.

Polcaro, A.M., Palmas, S., Renoldi, F. & Mascia, M. (2000) Three-dimensional electrodes for the electrochemical combustion of organic pollutants, *Electrochimica Acta* 46(2-3), the International Society of Electrochemistry (ISE), Elsevier, 389-394.

Polcaro, A.M., Vacca, A., Mascia, M. & Palmas, S. (2005) Oxidation at boron doped diamond electrodes: An effective method to mineralize triazines, *Electrochimica Acta* 50(9), the International Society of Electrochemistry (ISE), Elsevier, 1841-1847.

Polcaro, A.M., Vacca, A., Palmas, S. & Mascia, M. (2003) Electrochemical treatment of wastewater containing phenolic compounds: Oxidation at boron-doped diamond electrodes, *Journal of Applied Electrochemistry* 33(10), Springer, 885-892.

Ponce-de-Leon, C., Pletcher, D. (1995) Removal of formaldehyde from aqueous solutions via oxygen reduction using a reticulated vitreous carbon cathode cell, *Journal of Applied Electrochemistry* 25(4), Springer, 307-314.

Pratap, K., Lemley, A.T. (1994) Electrochemical peroxide treatment f aqueous herbicide solutions, *Journal of Agricultural and Food Chemistry* 42(1), American Chemical Society (ACS), 209-215.

Pratap, K., Lemley, A.T. (1998) Fenton Electrochemical Treatment of Aqueous Atrazine and Metolachlor, *Journal of Agricultural and Food Chemistry* 46(8), American Chemical Society (ACS), 3285-3291.

Pulgarin, C., Adler, N., Peringer, P. & Comninellis, C. (1994) Electrochemical detoxification of a 1,4-benzoquinone solution in wastewater treatment, *Water Research* 28, International Water Association (IWA), 887-893.

Quiroz, M.A., Sanchez-Salas, J.L., Reyna, S., Bandala, E.R., Peralta-Hernandez, J.M. & Martinez-Huitle, C.A. (2014) Degradation of 1-hydroxy-2,4-dinitrobenzene from aqueous solutions by electrochemical oxidation: Role of anodic material, *Journal of Hazardous Materials* 268, Elsevier, 6-13.

Real, F.J., Benitez, F.J., Acero, J.L., Roldan, G. & Casas, F. (2012) Elimination of the Emerging Contaminants Amitriptyline Hydrochloride, Methyl Salicylate, and 2-Phenoxyethanol in Ultrapure Water and Secondary Effluents by Photolytic and Radicalary Pathways, *Journal of Industrial & Engineering Chemistry Research* 51(50), American Chemical Society (ACS), 16209-16215.

Rodgers, J.D., Jedral, W. & Bunce, N.J. (1999) Electrochemical oxidation of chlorinated phenols, *Environmental Science & Technology* 33(9), American Chemical Society (ACS), 1453-1457.

Rodrigo, M.A., Michaud, P.A., Duo, I., Panizza, M., Cerisola, G. & Comninellis, C. (2001) Oxidation of 4-Chlorophenol at boron-doped diamond electrodes for wastewater treatment, Journal of The Electrochemical Society (JES) 148(5), *the society for solid-state and electrochemical science and technology* (ECS), D60-D64.

Rodrigo, M.A., Oturan, N. & Oturan, M.A. (2014) Electrochemically Assisted Remediation of Pesticides in Soils and Water: A Review, Chemical Reviews 114(17), American Chemical Society (ACS), 8720-8745.

Roe, B.A., Lemley, A.T. (1997) Treatment of two insecticides in an electrochemical Fenton system, *Journal of Environmental Science and Health Part B* 32(2), Taylor & Francis, 261-281.

Rojas, M.R., Leung, C., Whitely, D., Zhu, Y., Arnold, R.G. & Saez, E. (2011) Advanced Oxidation of Trace Organics in Water by Hydrogen Peroxide Solar Photolysis, Journal of Industrial & Engineering Chemistry Research 50(22), *American Chemical Society* (ACS), 12479-12487.

Rosales, E., Sanroman, M.A. & Pazos, M. (2012) Application of central composite face-centered design and response surface methodology for the optimization of electro-Fenton decolorization of Azure B dye, *Journal of Environmental Science and Pollution Research* 19(5), Springer, 1738-1746.

Rossi, A.F., Martins, R.C. & Quinta-Ferreira, R.M. (2014) Composition Effect of Iron-Copper Composite Catalysts in the Fenton Heterogeneous Process Efficiency and Cooxidation Synergy Assessment, *Journal of Industrial & Engineering Chemistry Research,* American Chemical Society (ACS).

Rychen, P., Pupunat, L., Haenni, W. & Santoli, E. (2003) Water treatment applications with BDD electrodes and the DiaCell concept, *New Diamond and Frontier Carbon Technology* 13, MYU Publisher, 109-117.

Samet, Y., Elaoud, S.C., Ammar, S. & Abdelhedi, R. (2006) Electrochemical degradation of 4-chloroguaiacol for wastewater treatment using PbO_2 anodes, *Journal of Hazardous Materials* 138(3), Elsevier, 614-619.

Saracco, G., Solarino, L., Aigotti, R., Specchia, V. & Maja, M. (2000) Electrochemical oxidation of organic pollutants at low electrolyte concentrations, Electrochimica Acta 46(2-3), *International Society of Electrochemistry* (ISE), Elsevier, 373-380.

Schmidt-Szalowski, K., Krawczyk, K., Sentek, J., Ulejczyk, B., Gorska, A. & Mlotek, M. (2011) Hybrid plasma-catalytic systems for converting substances of high stability, greenhouse gases and VOC, *Chemical Engineering Research and Design* 89(12), Elsevier, 2643-2651.

Sharma, V.K., Millero, F.J. (1988) Oxidation of copper (I) in seawater, *Environmental science & technology* 22(7), 768-771.

Shih, Y.J., Lin, C.P. & Huang, Y.H. (2013) Application of Fered-Fenton and chemical precipitation for the treatment of electroless nickel plating wastewater, *Separation and Purification Technology* 104, Elsevier, 100-105.

Simond, O., Schaller, V. & Comninellis, C. (1997) Theoretical model for the anodic oxidation of organics on metal oxide electrodes, Electrochimica Acta 42, *International Society of Electrochemistry* (ISE), Elsevier, 2009-2012.

Sires, I., Brillas, E., Oturan, M.A., Rodrigo, M.A. & Panizza, M. (2014) Electrochemical advanced oxidation processes: today and tomorrow. A review, *Journal of Environmental Science and Pollution Research* 21(14), Springer, 8336-8367.

Socha, A., Chrzescijanska, E. & Kusmierek, E. (2005) Electrochemical and photoelectrochemical treatment of 1-aminonaphthalene-3,6-disulphonic acid, *Dyes and Pigments* 67(1), Elsevier, 71-75.

Song, L.J., Zhu, N.W., Yuan, H.P., Hong, Y. & Ding, J. (2010) Enhancement of waste activated sludge aerobic digestion by electrochemical pre-treatment, Water Research 44(15), *the International Water Association* (IWA), Elsevier, 4371-4378.

Song, S., Zhan, L., He, Z., Lin, L., Tu, J., Zhang, Z., Chen, J. & Xu, L. (2010) Mechanism of the anodic oxidation of 4-chloro-3-methyl phenol in aqueous solution using Ti/SnO_2-Sb/PbO_2 electrodes, *Journal of Hazardous Materials* 175(1-3), Elsevier, 614-621.

Soriaga, M.P., Hubbard, A.T. (1982) Determination of the orientation of adsorbed molecules at solid-liquid interfaces by thin-layer electrochemistry: Aromatic compounds at platinum electrodes, *Journal of American Chemical Society* 104, 2735-2742.

Stucki, S., Koetz, R., Carcer, B. & Suter, W. (1991) Electrochemical wastewater treatment using high overvoltage anodes. Part II: Anode performance and applications, *Journal of Applied Electrochemistry* 21(2), Springer, 99-104.

Subba Rao, A.N., Venkatarangaiah, V.T. (2014) Metal oxide-coated anodes in wastewater treatment, *Journal of Environmental Science and Pollution* Research 21(5), Springer, 3197-3217.

Sudoh, M., Kitaguchi, H. & Koide, K.J. (1985) Polarization Characteristics of Packed Bed Electrode Reactor for Electroreduction of Oxygen to Hydrogen Peroxide, Journal of Chemical Engineering of Japan 18(4), *The Society of Chemical Engineers Japan* (SCEJ), 364-371.

Sudoh, M., Kitaguchi, H. & Koide, K.J. (1985) Electrochemical Production of Hydrogen Peroxide by Reduction of Oxygen, Journal of Chemical Engineering of Japan 18(5), *The Society of Chemical Engineers Japan* (SCEJ), 409-414.

Sugimoto, W., Yokoshima, K., Murakami, Y. & Takasu, Y. (2006) Charge storage mechanism of nanostructured anhydrous and hydrous ruthenium-based oxides, Electrochimica Acta 52(4), *International Society of Electrochemistry* (ISE), Elsevier, 1742-1748.

Szpyrkowicz, L., Juzzolino, C., Kaul, S.N., Daniele, S. & De Faveri, M. (2000) Electrochemical Oxidation of Dyeing Baths Bearing Dispers Dyes, Journal of Industrial & Engineering Chemistry Research 39(9), *American Chemical Society* (ACS), 3241-3248.

Tran, L.H., Drogui, P., Mercier, G. & Blais, J.F. (2009) Electrochemical degradation of polycyclic aromatic hydrocarbons in creosote solution using ruthenium oxide an titanium expanded mesh anode, *Journal of Hazardous Materials* 164(2-3), Elsevier, 1118-1129.

Tarr, M.A. (2003) Fenton and Modified Fenton Methods for Pollutant Degradation, In: Chemical Degradation Methods for Wastes and Pollutants, *Environmental and Industrial Applications,* Part 4, Edited by M.A. Tarr, Marcel Dekker Inc., New York-Basel.

Troster, I., Fryda, M., Herrmann, D., Schafer, L., Haenni, W., Perret, A., Blaschke, M., Kraft, A. & Stadelmann, M. (2002) Electrochemical advanced oxidation process for water treatment using DiaChem® electrodes, *Diamond and Related Materials* 11(3-6), Elsevier, 640-645.

Vandenbroucke, A.M., Morent, R., De Geyter, N. & Leys, C. (2011) Non-thermal plasmas for non-catalytic and catalytic VOC abatement, *Journal of Hazardous Materials* 195, Elsevier, 30-54.

Vargas, R., Borras, C., Mostany, J. & Scharifker, B.R. (2010) Measurement of phenols dearomatization via electrolysis: The UV-Vis solid phase extraction method, Water Research 44(3), *the International Water Association* (IWA), Elsevier, 911-917.

Vicent, F., Morallon, E., Quijada, C., Vazquez, J.L., Aldaz, A. & Cases, F. (1998) Characterization and stability of doped SnO_2 anodes, *Journal of Applied Electrochemistry* 28(6), Springer, 607-612.

Vijayalakshmi, P., Bhaskar Raju, G. & Gnanamani, A. (2011) Advanced Oxidation and Elektrooxidation As Tertiary Treatment Techniques to Improve the Purity of Tannery Wastewater, Journal of Industrial & Engineering Chemistry Research 50(17*), American Chemical Society* (ACS), 10194-10200.

Wang, H.J., Chen, X.Y. (2011) Kinetic analysis and energy efficiency of phenol degradation in a plasma-photocatalysis system, *Journal of Hazardous Materials* 186(2-3), Elsevier, 1888-1892.

Wang, Y.R., Chu, W. (2011) Degradation of 2,4,5-trichlorophenoxyacetic acid by a novel Electro-Fe(II)/Oxone process using iron sheet as the sacrificial anode, Water Research 45(13), *International Water Association* (IWA), 3883-3889.

Wang, X.Q., Liu, C.P., Yuan, Y. & Li, F.B. (2014) Arsenite oxidation and removal driven by a bio-electro-Fenton process under neutral pH conditions, *Journal of Hazardous Materials* 275, Elsevier, 200-209.

Wang, A., Qu, J., Ru, J., Liu, H., & Ge, J. (2005) Mineralization of an azo dye Acid Red 14 by electro-Fenton's reagent using an activated carbon fiber cathode, *Dyes and Pigments* 65(3), Elsevier, 227-233.

Wang, Y., Shen, Z. & Chen, X. (2010) Effects of experimental parameters on 2,4-dichlorphenol degradation over Er-chitosan-PbO_2 electrode, *Journal of Hazardous Materials* 178(1-3), Elsevier, 867-874.

Wang, H., Wang, J. (2007) Electrochemical degradation of 4-chlorophenol using a novel Pd/C gas-diffusion electrode, *Applied Catalysis B: Environmental* 77(1-2), Elsevier, 58-65.

Wang, H., Yu, X., Wu, L., Wang, Q. & Sun, D. (2005) A Comparative Study on Electrochemical Oxidation of Phenol in Two Types of Cells, *Russian Journal of Electrochemistry* 41(7), Springer, 719-724.

Wang, Y., Zhao, G., Chai, S., Zhao, H. & Wang, Y. (2013) Three-Dimensional Homogeneous Ferrite-Carbon Aerogel: One Pot Fabrication and Enhanced Electro-Fenton Reactivity, *Applied Materials & Interfaces 5(3), American Chemical Society,* 842-852.

Wang, Y., Zhao, H., Gao, J., Zhao, G., Zhang, Y. & Zhang, Y. (2012) Rapid Mineralization of Azo-Dye Wastewater by Microwave Synergistic Electro-Fenton Oxidation Process, The Journal of Physical Chemistry C 116(13), *American Chemical Society*, 7457-7463.

Weiss, E., Groenen-Serrano, K. & Savall, A. (2008) A comparison of electrochemical degradation of phenol on boron doped diamond and lead dioxide anodes, *Journal of Applied Electrochemistry* 38(3), Springer, 329-337.

Woisetschlaeger, D., Humpl, B., Koncar, M. & Siebenhofer, M. (2013) Electrochemical Oxidation of Wastewater – opportunities and drawbacks, *Water Science & Technology,* IWA Publishing.

Wu, X., Zhang, H., Li, Y., Zhang, D. & Li, X. (2014) Factorial design analysis for COD removal from landfill leachate by photoassisted Fered-Fenton process, *Journal of Environmental Science and Pollution Research* 21(14), Springer, 8595-8602.

Xu, L., Guo, Z. & Du, L. (2013) Anodic oxidation of azo dye C.I.Acid Red 73 by the yttrium-doped Ti/SnO_2-Sb electrodes, *Frontiers of Chemical Science and Engineering* 7(3), Springer, 338-346.

Yang, X., Zou, R., Huo, F., Cai, D. & Xiao, D. (2009) Preparation and characterization of Ti/SnO_2-Sb_2O_3-Nb_2O_5/PbO_2 thin film as electrode material for the degradation of phenol, *Journal of Hazardous Materials* 164(1), Elsevier, 367-373.

Yao, Y., Zhao, C., Zhao, M. & Wang, X. (2013) Electrocatalytic degradation of methylene blue on PbO_2-ZrO_2 nanocomposite electrodes prepared by pulse electrodeposition, *Journal of Hazardous Materials* 263(2), Elsevier, 726-734.

Yavuz, Y., Shahbazi, R., Koparal, A.S. & Oeguetveren, Ue.B. (2014) Treatment of Basic Red 29 dye solution using iron-aluminum electrode pairs by electrocoagulation and electro-Fenton methods, *Journal of Environmental Science and Pollution Research* 21(14), Springer, 8603-8609.

Yu, F., Zhou, M., Zhou, L. & Peng, R. (2014) A Novel Electro-Fenton Process with H_2O_2 Generation in a Rotating Disk Reactor for Organic Pollutant Degradation, Environmental Science & Technology Letters 1(7), *American Chemical Society* (ACS), 320-324.

Yuan, C., Chen, C.Y. & Huang, C.H. (2013) Electrochemical remediation of BPA in a soil matrix by Pb/Ti and RuO_2/Ti electrodes, *Journal of Applied* Electrochemistry 43(12), Springer, 1163-1174.

Yuan, S., Fan, Y., Zhang, Y., Tong, M. & Liao, P. (2011) Pd-Catalytic In Situ Generation of H_2O_2 from H_2 and O_2 Produced by Water Electrolysis for the Efficient Electro-Fenton Degradation of Rhodamine B, Environmental Science & Technology 45(19), *American Chemical Society* (ACS), 8514-8520.

Yu, R.F., Lin, C.H., Chen, H.W., Cheng, W.P. & Kao, M.C. (2013) Possible control approaches of the Electro-Fenton process for textile wastewater treatment using on-line monitoring of DO and ORP, *Chemical Engineering Journal* 218, Elsevier, 341-349.

Zanta, C.L.P.S., De Andrade, A.R. & Boodts, J.F.C. (2000) Electrochemical behavior of olefins: Oxidation at ruthenium-titanium dioxide and iridium-titanium dioxide coated electrodes, *Journal of Applied Electrochemistry* 30(4), Springer, 467-474.

Zarei, M., Niaei, A., Salari, D. & Khataee, A. (2010) Application of response surface methodology for optimization pf peroxi-coagulation of textile dye solution using carbon nanotube-PTFE cathode, *Journal of Hazardous Materials* 173(1-3), Elsevier, 544-551.

Zhang, G., Wang, S. & Yang, F. (2012) Efficient Adsorption and Combined Heterogeneous/Homogeneous Fenton Oxidation of Amaranth Using supported Nano-FeOOH as Cathodic Catalysts, The Journal of Physical Chemistry C 116(5*), American Chemical Society* (ACS), 3623-3634.

Zhang, G., Yang, F., Gao, M. & Liu, L. (2008) Electrolytic Behavior of the Bare and the Anthraquinonedisulfonate/Polypyrrole Composite Film Modified Graphite Cathodes in the Electro-Fenton System, The Journal of Physical Chemistry C 112(24), *American Chemical Society (*ACS), 8957-8962.

Zhang, H., Wu, X. & Li, X. (2012) Oxidation and coagulation removal of COD from landfill leachate by Fered-Fenton process, *Chemical Engineering Journal* 210, Elsevier, 188-194.

Zhang, J., Zheng, Z., Zhang, Y., Feng, J. & Li, J. (2008) Low-temperature plasma-induced degradation of aqueous 2,4-dinitrophenol, *Journal of Hazardous Materials 154*(1), Elsevier, 506-512.

Zhang, M., Yao, Q., Guan, W., Lu, C. & Lin, J.M. (2014) Layered Double Hydroxide-Supported Carbon Dots as an Efficient Heterogeneous Fenton-Like Catalyst for Generation of Hydroxyl Radicals, The Journal of Physical Chemistry C 118(19), *American Chemical Society* (ACS), 10441-10447.

Zhang, T., Chen, Y., Wang, Y., LeRoux, J., Yang, Y. & Croue, J.P. (2014) Efficient Peroxydisulfate Activation Process Not Relying on Sulfate Radical Generation for Water Pollutant Degradation, Environmental Science & Technology 48(10), *American Chemical Society* (ACS), 5868-5875.

Zhang, Y.M., Peng, Y., Yin, X.L., Liu, Z.H. & Li, G. (2014) Degradation of lignin to BHT by electrochemical catalysis on Pb/PbO_2 anode in alkaline solution, *Journal of Chemical Technology and Biotechnology, Society of Chemical Industry* (SCI), Wiley.

Zhao, G., Zhang, Y., Lei, Y., Lv, B., Gao, J., Zhang, Y. & Li, D. (2010) Fabrication and Electrochemical Treatment Application of A Novel Lead Dioxide Anode with Superhydrophobic Surfaces, High Oxygen Evolution Potential, and Oxidation Capability, Environmental Science & Technology 44(5), *American Chemical Society* (ACS), 1754-1759.

Zhou, M., Dai, Q., Lei, L., Ma, C. & Wang, D. (2005) Long Life Modified Lead Dioxide Anode for Organic Wastewater Treatment : Electrochemical Characteristics and Degradation Mechanism, Environmental Science & Technology 39(1), *American Chemical Society* (ACS), 363-370.

Zhou, M., He, J. (2008) Degradation of cationic red X-GRL by electrochemical oxidation on modified PbO_2 electrode, *Journal of Hazardous Materials 153(1-2), Elsevier, 357-363.*

In: Enhanced Electrochemical Advanced Oxidation ... ISBN: 978-1-63482-089-9
Editor: Gagik Badalians Gholikandi

Chapter 4

WASTE ACTIVATED SLUDGE STABILIZATION AND REUSE EMPLOYING FERED-FENTON TECHNOLOGY

Gagik Badalians Gholikandi*, Shafieh Karami, Mina Nili Ardakani and Hamidreza Masihi
Faculty of Water Engineering and Environment,
Shahid Beheshti University, A.C., Tehran, Iran

INTRODUCTION

Treatment and safe disposal of hazardous materials (waste, excess sludge, leachate, etc.) is an imperative and inevitable measure to preserve environment. Nowadays, wastewater treatment systems are used widely by activated sludge method, which results in producing a large amount of biological excess sludge (Tokumura et al., 2009; Xing et al., 2011). Increasing growth of the biological excess sludge has made it one of the most important and critical environmental issues (Li et al., 2011; Xu et al., 2012). Effective sludge management is considered as key and strategic goal in wastewater treatment systems (Samaras et al., 2008; Yu et al., 2010). The waste sludge contains hazardous substances, such as chemical organic materials, heavy metals and pathogens. Therefore, if it is not stabilized and disposed safely, they will enter the food chain and endanger the human health respectively (Wang et al., 2012; Wang et al., 2013; Xu et al., 2013). The overall objective of stabilizing the waste sludge includes removal of organic matter, heavy metals, and pathogens, and the sludge volume reduction, converting them into useful sources (Carrère et al., 2010; Weng et al., 2014).

Recently, the use of advanced oxidation processes (AOPs), especially the electrochemical advanced oxidation processes (EAOPs) has been taken into consideration in wastewater treatment (Gu et al., 2013). Performance efficiency of advanced oxidation processes in degradation of biological organic compounds, wastewater sludge contaminants and soil has caused attractiveness of such methods (He and Wei 2010; Shen et al., 2013). Advanced oxidation processes are those in which adequate hydroxyl radical is generated (Anotai et al.,

* Corresponding Author, Hakimieh- Tehranpars, Shahid Abbaspour Blvd., PO Code 16589-53571, PO Box 16765-1719, Tehran, Iran. Email: g.badalians@yahoo.com

2010; Liu et al., 2012). Hydroxyl radical which is one of the most impressive oxidizing agents in water and environment, reacts non-selectively (Gu et al., 2013; Rabaaoui et al., 2013) ($E^0 = 2.87$) (Tao et al., 2013). One of the most common advanced oxidation methods is the Fenton process that has advantages such as low initial cost, easy operation and production of non-toxic byproducts (Anotai et al., 2010). In this study, combined Fenton and electrochemical method, i.e., Fered-Fenton reactor was employed in order to reduce the waste activated sludge organic load, i.e.Volatile suspended solids (VSS). The consumption rate of divalent iron ions in the Fenton process is higher than its reduction (the reaction 2). It can be considered as a noticeable disadvantage (Zhang et al., 2006; Babuponnusami and Muthukumar 2012). To overcome this issue, the electrochemical method was employed. The use of this method causes an increased reduction rate of divalent iron by converting the trivalent iron ions to the divalent ones (the reaction3) (Badalians Gholikandi et al., 2014).

$$Fe^{2+} + H_2O_2 \rightarrow Fe^{3+} + OH^0 + OH^- \qquad 53-76M^{-1}S^{-1} \tag{1}$$

$$Fe^{3+} + H_2O_2 \rightarrow Fe^{2+} + {}^0H_2O + H^+ \qquad 0.01-0.02M^{-1}S^{-1} \tag{2}$$

$$Fe^{3+} + e^- \rightarrow Fe^{2+} \tag{3}$$

In this study, the related investigations were conducted in three stages, i.e., determination of organics loading removal efficiency of the Fered-Fenton reactor, determination of the reactor's main specifications under optimum conditions, and finally, determination of the reactor's capability to remove heavy metals contained in the sludge and to improve the sludge characteristics relating to the reuse purposes.

EXPERIMENTAL INVESTIGATIONS

The investigations were conducted in a laboratory scale pilot. The lab-scale Fered-Fenton reactor has been shown in Figure1. The reactor was made of Plexiglass with a volume equals 0.9 L. Because of the chemical inert and decomposable property of graphite, two graphite anodes and two graphite cathodes were employed. The electrodes dimension were 1 mm× 140 mm× 60 mm, and the distance between them was 1.5 cm (Nidheesh et al., 2012). The depth of the electrodes placed in the sludge was 100 mm and its contact surface was 100 mm× 60 mm. An electrical mixer (Zheng, zs-ri, 6(V), DC, 3.66 rpm) was employed accordingly. A digital laboratory model power supply (Mps, DC-3003D, 0-3 (A), 0-30 (V)) has been used for the amperage adjustment (figure 1). The used chemicals were ferrous sulfate ($FeSO_4$), hydrogen peroxide (H_2O_2), and concentrated sulfuric acid (H_2SO_4) from Merck® Co. Whatman® filter paper (no. 42) was also utilized. Waste activated sludge samples were taken from a municipal wastewater treatment plant in Tehran employing activated sludge/ extended aeration process. Table 1 shows the specifications of the characterized sludge (Badalians Gholikandi et al., 2014).

Table1. Waste activated Sludge Specifications

Property	Value
pH	6.53-7.33
Temperature (°C)	13.8-15.6
COD (mg/L)	6051-11164
VSS (mg/L)	3536-7782
VS (mg/L)	4086-8382
TSS (mg/L)	4088-8432
VSS/TSS	0.86-0.92

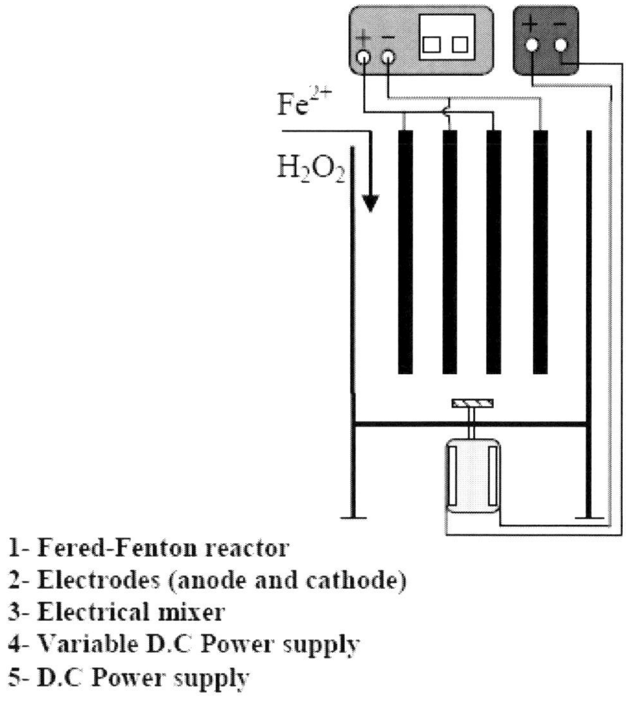

Figure 1. The Fered-Fenton reactor in lab-scale.

The pH value of samples were adjusted by adding an adequate dosage of Sulphuric acid and soda (NaOH) as required before ferrous sulfate and hydrogen peroxide are added to the system. Plates working as electrodes were connected to DC power supply apparatus and installed in the cell and then, voltage and current were adjusted. After hourly sampling from the depth of 100 mm of cell, VSS, and amperage are measured. All Sludge samples were analyzed by analytical methods referred to standard methods for the examination of water and wastewater (American Public Health Association, APHA, 1999). However, all experiments were repeated three times in order to minimize possible errors.

1. First Experimental Stage

In the first stage of the experiments, the VSS removal rate and efficiency, and various affecting parameters were investigated considering pH, the $[Fe^{2+}]/[H_2O_2]$ ratio, retention time, hydrogen peroxide concentration, and current density, and their relationship. Consequently, the optimum operational conditions have been determined. In addition, the results of a comparative study regarding the Fenton and Fered-Fenton processes occurrence have been presented.

$[Fe^{2+}]/[H_2O_2]$ ratio: The ratio of VSS removal efficiency in the absence of Fe^{2+} ions does not change and equals to 0. An excess of Fe^{2+} ions leads to hydroxyl radical generation and subsequently oxidation of organic matters (the reaction 1). Increasing ratio of $[Fe^{2+}]/[H_2O_2]$ from 0 to 0.145 gives rise to the efficiency of the system (equals to 35% VSS removal) and finally, the maximum efficiency of the system has been achieved by $[Fe^{2+}]/[H_2O_2]=0.58$ (equals to 70 % process efficiency). It can be observed that system efficiency does not show a considerable change due to the further increase of this ratio up to 0.87. Excessive concentrations of Fe^{2+} result in hydroxyl radical consumption which will reduce the efficiency of the system (the reaction 4) (Narayanan et al., 2003). Increasing ratio of $[Fe^{2+}]/[H_2O_2]$ to 1.015 leads to the system efficiency decrease by 1%. Finally, increasing ratio of $[Fe^{2+}]/[H_2O_2]$ to 1.45 causes a decreasing process efficiency (equals to 59% process efficiency) (figure 2).

$$OH^0 + Fe^{2+} \rightarrow Fe^{3+} + OH^- \quad (4)$$

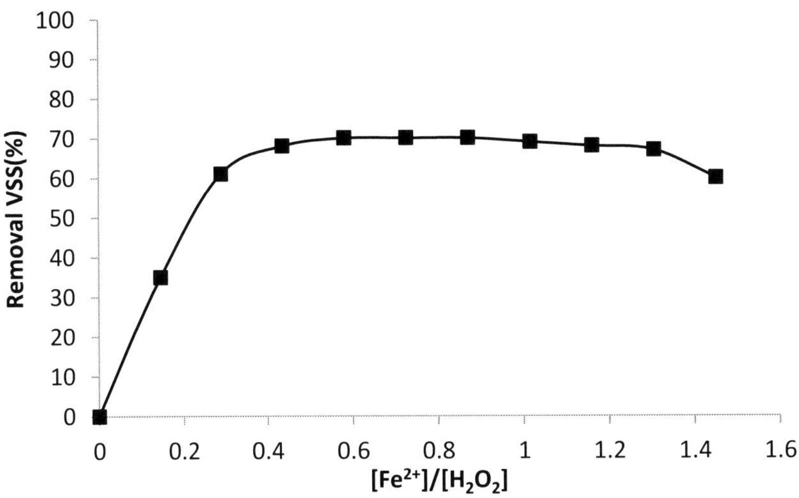

Figure 2. The VSS removal and $[Fe^{2+}]/[H_2O_2]$ ratios relationship (pH = 3, $[H_2O_2]$ = 1255 (mg/L), *I* = 600 (mA), Time = 4 (hr)).

pH value: The results of the previous studies have determined that the optimum pH value for the Fenton process occurrence in wastewater treatment equals to about 3 (Ai et al., 2007; Zhou et al., 2007; Wang et al., 2008; Ghoneim et al., 2011). The pH value is one of the important affecting process factors. The experimental results (figure 3) showed that the maximum efficiency of VSS removal occurred at pH 3.1 (the removal efficiency equals to

73%). Furthermore, the decreasing amount of pH to less than 3 is coincident with the reducing process efficiency. However, while pH value is equal to 1, the process efficiency is equal to 62%. Essentially, the oxidation efficiency decreases while pH values are lower than 3 because of the resulting stable complexes from Fe^{2+} ions and H_2O_2 which leads to deactivation of Fe^{2+} as a catalyst (Wang et al., 2010); additionally, the deactivated hydrogen peroxide in the form of $H_3O_2^+$ (Zhou et al., 2007) ends in not participating of H_2O_2 in the reaction 1, a reduction of the hydroxyl radical generation, and consequently, reducing VSS removal efficiency. Additionally, if the pH value increases from 3 to 5, the process efficiency declines continuously. At the pH value equals to 5, the VSS removal efficiency was 67%. However, at pH ranges above 5, as hydrogen peroxide is unstable and decomposes into water and oxygen, the efficiency of the system decline considerably. VSS removal efficiency decrease at pH values above 7 significantly (Wang et al. 2010; Shemer et al. 2006). According to US EPA, the maximum VSS removal efficiency for the conventional sludge stabilization methods is about 50% (US EPA 1995). The experimental results show that the VSS removal efficiency at pH values of 5 to 7 (pH values common range under usual operation conditions) is more than 50% and therefore, the neutralization process is not a necessary stage.

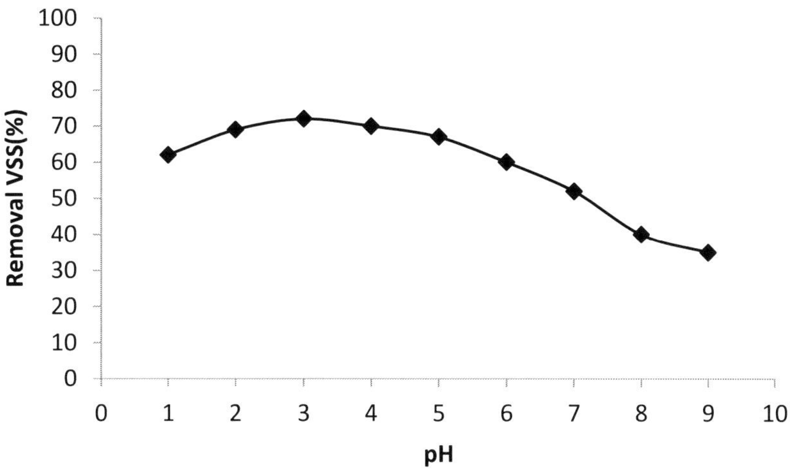

Figure 3. The VSS removal rates as a function of the pH values ($[Fe^{2+}]/[H_2O_2] = 0.58$, $[H_2O_2] = 1255$ (mg/L), I =600 (mA), Time = 4 (hr)).

Retention time: It is important to attend Retention time as a significant operational parameter. In the most cases of conventional sludge stabilization methods, i.e., aerobic and anaerobic digestion, composting, the retention time is considered as several days. But in Fered-Fenton system it is considerably reduced. The reduced retention time facilitates operational activities and also considerably lessens reactor volume. As shown in Figure 4, in the two-hour preliminary experiments, the most part of the removal efficiency, i.e., equals to about 50%, could be obtained. It is due to the high concentration of degradable organics in the system. After 2 hours, the efficiency of the system equals to 70% and after 6 hours increases to 76%. It is important to mark that the retention time is a considerable parameter from the energy point of view.

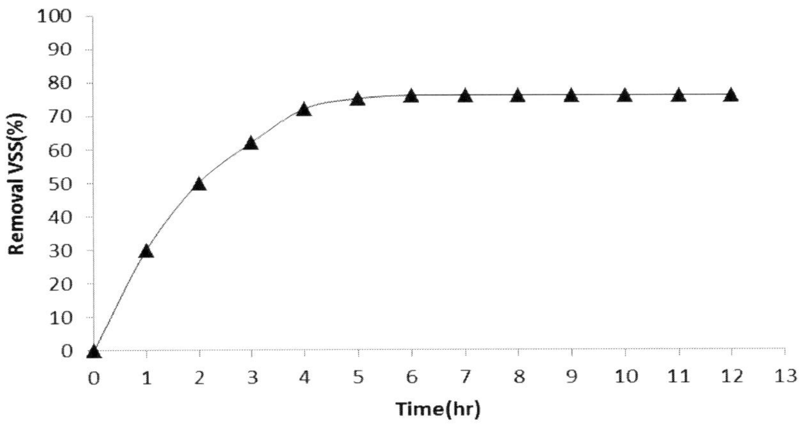

Figure 4. The VSS removal and time relationship ($[Fe^{2+}]/[H_2O_2]$ = 0.58, pH = 3.1, $[H_2O_2]$ =1255 (mg/L), I = 600 (mA)).

Hydrogen peroxide concentration: The VSS removal efficiency in the absence of the hydrogen peroxide equals to 0% as shown in Figure 5. Increasing in hydrogen peroxide concentration leads to hydroxyl radical generation (according to the reaction 1), and subsequently, the VSS reduction can be observed. It is equal to the maximum removal efficiency of 78% at the concentration of 1.568 mg/L. However, the more the concentration of hydrogen peroxide increases, the more the efficiency of the laboratory setup decreases. This is due to the consumption of the hydroxyl radicals that is evident in the reactions 5 and 6 (Muruganandham and Swaminathan 2004).

$$OH^0 + H_2O_2 \rightarrow HO_2^0 + H_2O \tag{5}$$

$$OH^0 + HO_2^0 + \rightarrow H_2O + O_2 \tag{6}$$

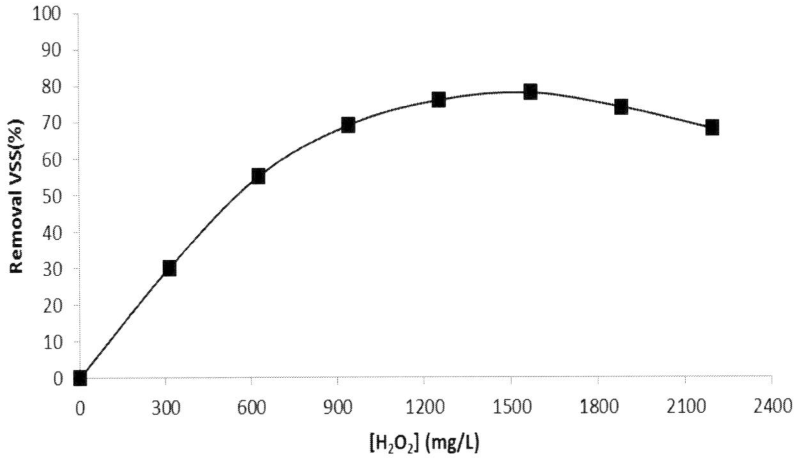

Figure 5. The VSS removal and $[H_2O_2]$ relationship($[Fe^{2+}]/[H_2O_2]$ = 0.58, pH = 3.1, I = 600 (mA), Time = 6 (hr)).

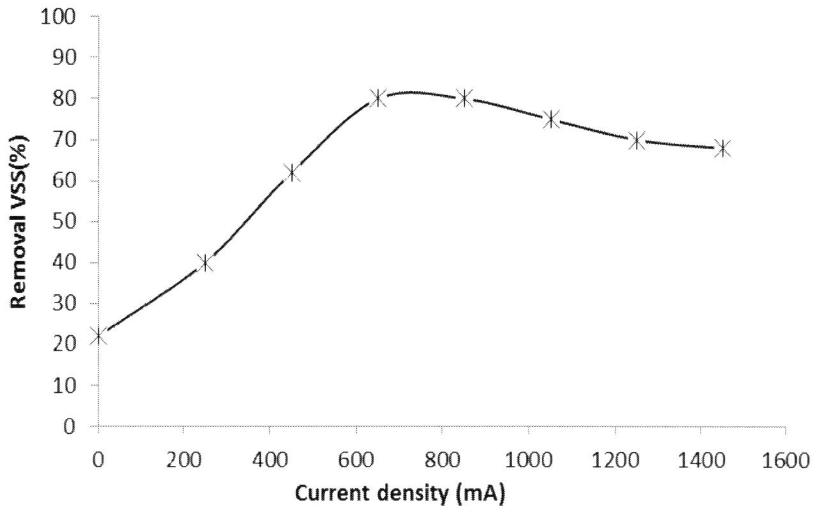

Figure 6. The VSS removal and current density relationship ($[Fe^{2+}]/[H_2O_2] = 0.58$, pH = 3.1, $[H_2O_2] =$ 1568 (mg/L), Time = 6 (hr)).

Current density: Increasing in current density ameliorates reduction of Fe^{3+} ions to Fe^{2+} that resulting in an enhancement in Fenton process efficiency and the reaction 3 occurs (Muruganandham and Swaminathan, 2004). It is obvious from Figure 6 that there is only Fenton process while no current in the system is established (process efficiency equals to 22%). In the current interval from 650 to 850 mA, the maximum VSS removal efficiency of 81% is achieved. Higher current density has an inhibiting effect on the reaction 3 and declines the efficiency of the process due to the evolution of hydrogen at cathode (Zhang et al., 2007). As a result, in current density equals to 1,250 mA, the system efficiency decreases about 7%.

The Comparison between Fenton and Fered-Fenton Process

Degradation rate of organic matter in experimental lab-scale reactor is higher than using Fenton process singly. So, a comparison between Fenton and Fered-Fenton process shows that the VSS removal efficiency equals to 76% after 6h in combined method while, it is about 22% in Fenton process. This is due to the Fe^{2+} ions consumption for hydroxyl radical production in a higher rate than its reduction in Fenton process. The related reactions (the reactions 1and 2) (Babuponnusami et al., 2012; Su et al., 2012) show that the rate of the Fe^{2+} ions generation is less than its consumption and consequently, the production of iron ions as catalyst of the hydroxyl radical decreases. This doesn't occur in the Fered-Fenton method, due to the renewed reduction of iron ions using the electrochemical reaction 3. Therefore, the reaction occurs comparatively in a higher rate. In addition, because of using hydrogen peroxide to Fe^{3+} reduction in the Fenton process (the reaction 2), the efficiency of Fenton process is lower than the combined method in equal concentrations of the hydrogen peroxide (figure 7).

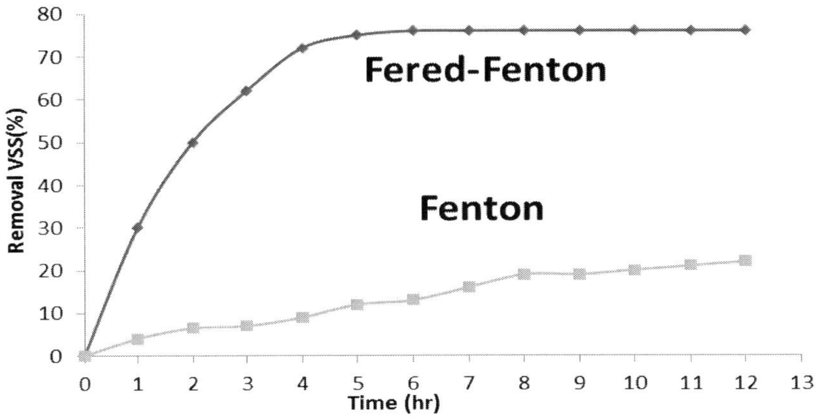

Figure 7. The VSS removal efficiency of Fenton and Fered-Fenton processes regarding the detention time ($[Fe^{2+}]/[H_2O_2] = 0.58$, pH = 3.1, $[H_2O_2]$ = 1568 (mg/L), I = 650 (mA).

2. Second Experimental Stage

Considering the previous research results, the optimum operational conditions of the pilot were determined as follows: pH=3, $[Fe^{2+}]/[H_2O_2]=0.58$, H_2O_2 concentration=1568 mg/L, current density=2.4 $mA/cm^2{}_{electrode}.L$, and the retention time equals to 4 hours. The experiments were conducted at a constant temperature, i.e., equals to 25 °C.

In this research stage, for optimizing the process and upgrading the efficiency of the Fered-Fenton reactor regarding waste activated sludge stabilization, the affecting parameters including inflow organics concentration, electrodes material, i.e., stainless steel, graphite, and iron, inter-electrode distance, number of Fenton's reagent injections, sodium sulfate concentration (the factor of the electrical conductivity generation in the medium), and the ratio of the electrodes surface to system volume were investigated.

Inflow organics concentration: As shown in the figure 8, increasing concentrations of the inflow organics results in rising VSS removal efficiency. However, the related upward rising curve possess a gentle slope; thus only a four-percent rise is obtained in the concentrations range of 3500-5000 mg/L. The feasible reason is the participation of the produced hydroxyl radical in the following reactions 7, 8, 9, and 10, as well as organic compounds oxidation (the reaction 11).

$$OH^0 + Fe^{2+} \rightarrow Fe^{3+} + OH^- \tag{7}$$

$$OH^0 + SO_4^{2-} \leftrightarrow OH^- + SO_4^{-0} \tag{8}$$

$$OH^0 + H_2O_2 \leftrightarrow OH_2^0 + H_2O \tag{9}$$

$$OH^0 + OH_2^0 \leftrightarrow +H_2O + O_2 \tag{10}$$

$$R^n + OH^0 \rightarrow R^{n-1} + OH^- \quad (11)$$

In other words, the whole hydroxyl radicals are not only dedicated to organics oxidation, but also some of them oxidize other ions available in the medium such as ferrous, sulfate, hydrogen peroxide and hydroperoxyl radical.

As the inflow organics concentration increases, the ratio of organics concentrations to hydroxyl radicals increases too, while the ratio of ferrous in the reaction 7, sulfate in the reaction 8, hydrogen peroxide in the reaction 9, and hydroperoxyl radical in reaction 10 to hydroxyl radical is constant. Therefore, increasing inflow organics concentration results in more hydroxyl radicals consumed to oxidize organics in the reaction 11 according to the stoichiometric coefficients. Consequently, the system efficiency increases.

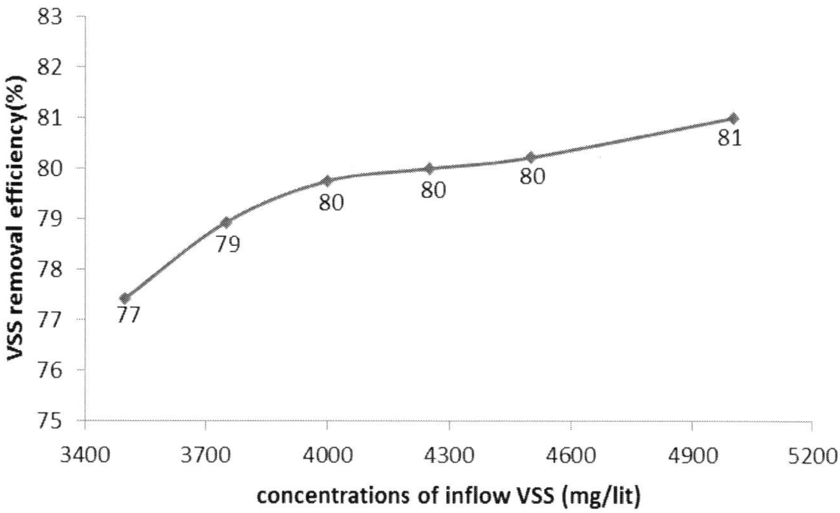

Figure 8. The relation between the VSS removal efficiency and the inflow organics concentration (pH = 3.1, $[Fe^{2+}]/[H_2O_2] = 0.58$, the current intensity = 650 mA, the retention time = 4(hr), the hydrogen peroxide concentration = 1568 mg/L, the electrodes material: graphite, the interelectrode distance= 1.5 cm, the number of Fenton's reagent injection=1, the sodium sulfate concentration = 0 mole/L, the ratio of electrodes surface to reactor volume= 266 cm^2/L).

Electrodes Material: Considering availability and technical-economic feasibility, graphite, stainless steel and iron electrodes were selected. As shown in the figure 9, graphite and iron electrodes have the highest and lowest system efficiency, respectively.

The experiments were performed by adding 909.44 mg/L ferrous ion to obtain the optimum ratio of ferrous ion to hydrogen peroxide equal to 0.58. According to the reactions12, 13, and 14 (Daneshvar et al., 2007; Modirsahla et al., 2008), iron electrode release ferrous ion in the medium; consequently, the mentioned ratio increases.

$$\left[\begin{array}{l} 4Fe_{(S)} \rightarrow 4Fe^{2+}_{(aq)} + 8e^- \\ 4H^+_{(aq)} + 8e^- \rightarrow 4H_{2(g)} \end{array}\right.$$

$$\Downarrow \quad (12,13)$$

$$4Fe_{(S)} + 8H_{(aq)} \rightarrow 4Fe^{2+}_{(aq)} + 4H_{2_{(g)}} \quad (14)$$

The excessive amount of the ferrous ion in the medium leads to hydroxyl radical consumption (the reaction7) and consequently, the system efficiency decreases. Graphite electrode is totally neutral and despite iron, doesn't take participate in chemical reactions occurring in cell; thus, the ratio of ferrous ion to hydrogen peroxide remains optimized and consequently, the maximum system efficiency is obtained.

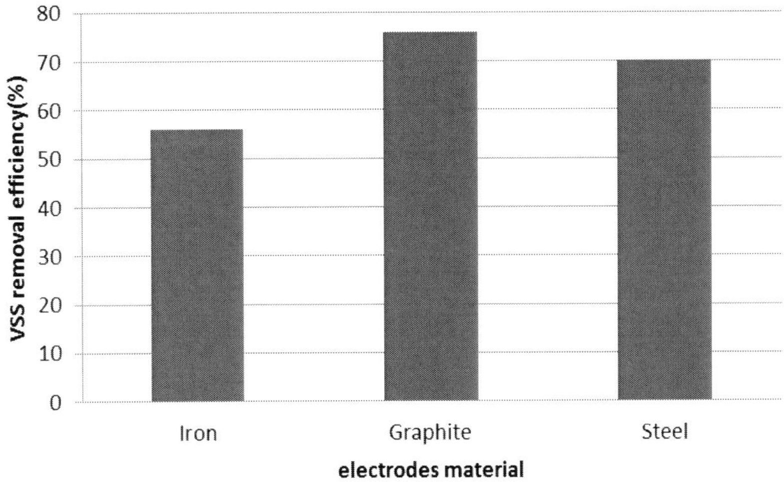

Figure 9. The relation between the VSS removal efficiency and the electrodes material (pH = 3.1, $[Fe^{2+}]/[H_2O_2] = 0.58$, the current intensity = 650 mA, the retention time = 4(hr), the hydrogen peroxide concentration = 1568 mg/L, the interelectrode distance = 1.5 cm, the number of Fenton's reagent injection = 1, the sodium sulfate concentration = 0 mole/L, the ratio of electrodes surface to reactor volume = 266 cm^2/L).

According to the electrodes materials selection, other affecting parameters, such as medium contamination, corrosion, operation and maintenance costs should be considered as well as the VSS removal efficiency. Graphite is an environmentally friendly material and leaves no residual in medium. Contrary to graphite, iron and stainless steel electrodes cause medium contamination. Stainless steel is more expensive than iron and graphite. Iron electrode has the drawback of being corroded in acidic medium and should be replaced after a short time. Graphite doesn't have corrosion problem and is appropriate for the long use; nevertheless since it is brittle and fragile, the carefully maintenance is required.

Inter-electrode distance: As shown in the figure 10, the effect of inter-electrode distance on the system efficiency corresponds with a quadratic equation; thus, as the distance increases from 0.5 to 1.5 cm, the system efficiency rises from 55% to 75% and the maximum VSS removal efficiency is obtained. If the distance increases and is more than 1.5 cm, the system efficiency declines and reaches 62%.

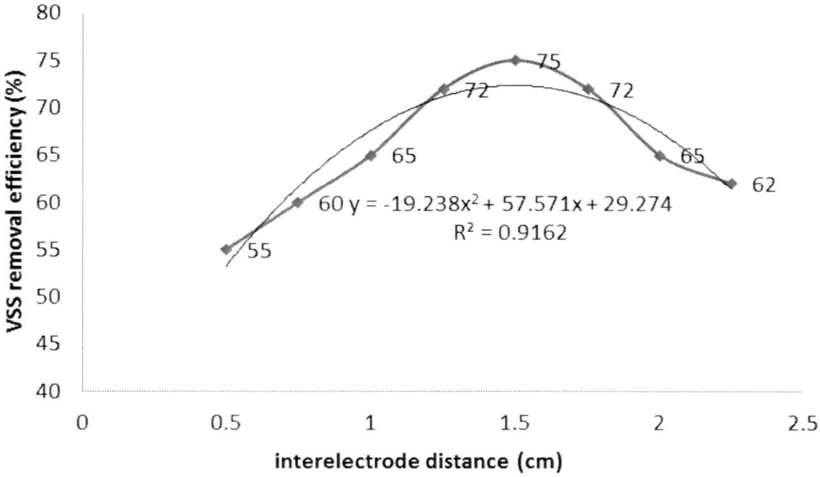

Figure 10. The relation between the VSS removal efficiency and the inter-electrode distance (pH = 3.1, $[Fe^{2+}]/[H_2O_2]$ = 0.58, the current intensity = 650 mA, the retention time = 4 (hr), the hydrogen peroxide concentration = 1568 mg/L, the electrodes material: graphite, the number of Fenton's reagent injection = 1, the sodium sulfate concentration = 0 mole/L, the ratio of the electrodes surface to the reactor volume = 266 cm^2/L).

Increasing inter-electrode distance leads to rising medium resistance. Therefore, according to the equation 15 and regarding the constant current density (650 mA), the power provided to the medium increases and as a result, more ferric ion is reduced to ferrous ion (the reaction 3) and the system efficiency is improved.

$$P = I^2 R \qquad (15)$$

where P is the power, R is the resistance, and I is the current density.

However, when the inter-electrode distance is increased more than a certain level equals to1.5 cm, the increase provides power causing the reaction 16 to reverse on one hand and the electrical coagulation occurrence following electrostatic attraction (Daneshvar et al., 2007) on the other hand. Consequently, the ferrous ion concentration in the medium is diminished based on the Fenton reaction 1, the hydroxyl radicals' generation for the oxidation of organic compounds decreases, and finally, the system efficiency will be lowered.

$$FeSO_4 \leftrightarrow Fe^{2+} + SO_4^{2-} \qquad (16)$$

Furthermore, the resistance rise following the distance increase makes generated ions to move more slowly that result in a decrease of the number of ferrous ions and hydrogen peroxide collision, hydroxyl radicals, and organic compounds. Finally, the system efficiency decreases accordingly.

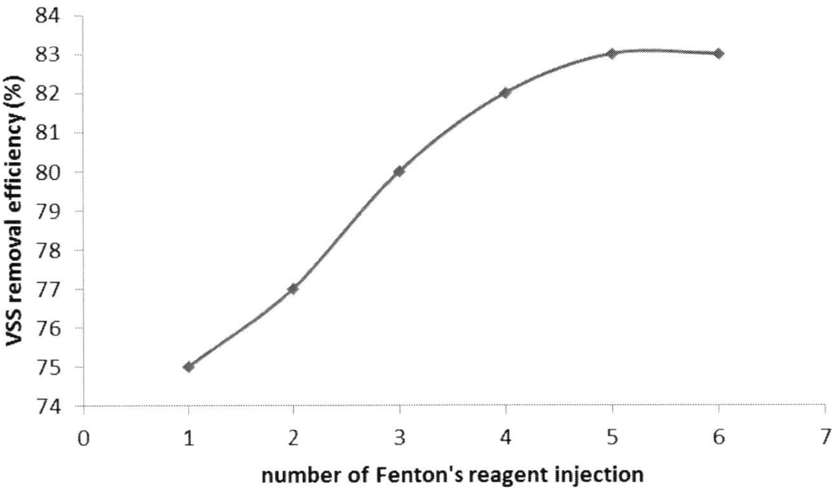

Figure 11. The relation between VSS removal efficiency and the number of Fenton's reagent injection (pH = 3.1, $[Fe^{2+}]/[H_2O_2]$ = 0.58, the current intensity = 650 mA, the retention time = 4 (hr), the hydrogen peroxide concentration = 1568 mg/L, the electrodes material: graphite, the inter-electrode distance = 1.5 cm, the sodium sulfate concentration = 0 mole/L, the ratio of electrodes surface to reactor volume = 266 cm^2/L).

Injection of the Fenton's reagent in stages: Considering the fact that the whole generated radicals are not allocated to the organics oxidation, some of them is consumed according to other ions present in the medium such as ferrous and sulfate, and also hydrogen peroxide and hydroperoxyl radical. In the first level of the multi-stage injection, the ratio of organics concentration to hydroxyl radical is more than that in single-stage injection. The organics concentration is similar in both methods. However, in each stage of the multi-stage injection method, hydroxyl radical concentration is less than in single-stage injection method, while the ratio of the ferrous ion in the reaction7, the sulfate in the reaction 8, the hydrogen peroxide in the reaction 9 and the hydroperoxyl radical in the reaction 10 to hydroxyl radical is the same under both conditions. Therefore, in each stage of the multi-stage injection method according to stoichiometric coefficients, more hydroxyl radicals' concentration is consumed for degrading organic compounds (the reaction 11).

As the number of the injection stages raises from 1 to 5, the system efficiency increases from 75% to 83% and at the stage 6 remains nearly constant (figure 11).

Electrode conductivity: Electrode conductivity is the moving capability of the ions in the medium which enhance the electron transfer rate. Since the electrolytes improve electrical conductivity, using them in solutions that are not conductive enough is required (Cheng-chun and Jia-fa 2007). As it is shown in the figure 12, the addition of sodium sulfate up to 0.111 moles/L results in increasing system efficiency equals to 88%. However, more addition conduces to a noticeable reduction in the efficiency.

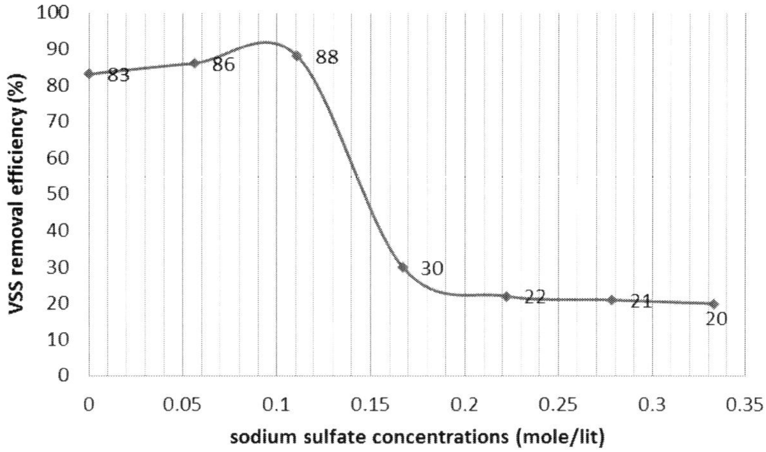

Figure 12. The relation between VSS removal efficiency and sodium sulfate concentration (pH = 3.1, $[Fe^{2+}]/[H_2O_2]$ = 0.58, the current intensity = 650 mA, the retention time = 4 (hr), the hydrogen peroxide concentration = 1568 mg/L, the electrodes material: graphite, interelectrode distance = 1.5 cm, the number of Fenton's reagent injections = 5, the ratio of electrodes surface to reactor volume = 266 cm^2/L).

When ferrous sulfate is added to the system following the hydrogen peroxide, the reactions 8, 16, and 17 occur in the medium, respectively (Eberson 1987; Zhou et al., 2007; Brillas et al., 2009).

$$SO_4^{-0} + e^- \leftrightarrow SO_4^{2-} \qquad E^0 = 2.6V / SHE \tag{17}$$

After providing the electrical current, the reaction 3 with E°=0.771 V/SHE takes place. Now adding sodium sulfate to the medium makes the occurrence of the reactions 8, 14, 18, and 19 possible.

$$Na_2SO_4 \leftrightarrow 2Na^+ SO_4^{2-} \tag{18}$$

$$Na^+ \leftrightarrow Na^{2+} + e^- \qquad E^0 - 2.714 \tag{19}$$

The reduction potentials mentioned above are possible under standard conditions (temperature = 25°C, concentration of ions = 1 molar and pressure = 1 atmosphere), otherwise Nernst equation (equation 20) is used to obtain the reduction potential (Stok et al., 1989).

$$E = E^0 + 2.3\frac{RT}{n,f}\log\frac{[a_{ax}]}{[a_{red}]} \tag{20}$$

where E° is standard reduction potential, R is universal constant of gases equal to 1.987.

Cal. K^{-1}. $mole^{-1}$ (Cal= Calorie) T is the absolute temperature in Kelvin, F is the Faraday constant equal to 23061 cal/V, n is the number of transferred electron, $[a_{ox}]$ and $[a_{red}]$ are the concentrations of oxidized and reduced ions in mole/L, respectively.

At room temperature (25 °C), RT/F may be treated like a constant and replaced by 25.693 mV, thus equation 21 is obtained (Stok et al., 1989).

$$E = E^0 + \frac{0.05916}{n}\log\frac{[a_{ax}]}{[a_{red}]} \tag{21}$$

Since the ions concentration in the solution is not 1 malar, the reduction potentials of sodium, sulfate and ferric, if adding 0.056 molar of sodium sulfate are respectively:

$$E = -2.714 + \frac{0.05916}{1}\log\frac{[2\times 0.56]}{[0.0163+0.0163+0.056]} = -2.708$$

$$E = 2.6 + \frac{0.05916}{1}\log\frac{[2\times 0.56]}{[0.0163+0.0163+0.056]} = 2.606$$

$$E = -0.771 + \frac{0.05916}{1}\log\frac{[2\times 0.56]}{[0.0163+0.0163+0.056]} = 0.777$$

Regarding following equation 22, the total reduction potential of the solution equals to 6.091.

$$E = E_{[a_{ax}]} - E_{[a_{red}]} \tag{22}$$

The calculations results related to the conditions when adding 0.111, 0.167, 0.222, 0.278, and 0.333 molar are recorded in the table 2.

Table 2. Reduction potential of the solution in different concentrations of sodium sulfate

Potential of solution (V/SHE)	Reduction potential of ferric ion (V/SHE)	Reduction potential of sulfate ion (V/SHE)	Reduction potential of sodium ion (V/SHE)	Sodium sulfate concentrations (mole/lit)
6.091	0.777	2.606	-2.708	0.056
6.096	0.782	2.611	-2.703	0.111
6.098	0.784	2.613	-2.701	0.167
6.099	0.785	2.614	-2.700	0.222
6.100	0.786	2.615	-2.699	0.278
6.100	0.786	2.615	-2.699	0.333

Regarding the above table, it's figured out that adding an initial amount of sodium sulfate equals to 0.056 molar causes an increasing reduction potential of the solution. Therefore, the electrochemical potential enhances and consequently, the reduction rate of ferric ion increases. It leads to raising system efficiency. When adding more sodium sulfate, the upswing trend in the reduction potential decreases gradually till it is finally steadied.

When the sodium sulfate concentration exceeds a certain amount (Nidheesh et al., 2012), the sulfate concentration augments in the solution. Consequently, it reacts with ferrous ion which was the necessary factor for the occurrence of the Fenton reaction 1 and precipitates (in other words, the reaction 16 occurs in opposite direction). On the other hand, it reacts with hydroxyl radical, producing sulfate radical anion ($SO_4^{-\circ}$) which has the weaker oxidation potential (Eberson 1987; Brillas et al., 2009), i.e., the oxidation potential of the medium decreases. While the sodium sulfate concentration increases more than 0.11 molar, there is a slight rise in the reduction potential. As a result, the negative effect of the decreasing ferrous ion concentration and medium oxidation potential dominates the positive effect of the increasing medium reduction potential. Consequently, the system efficiency decreases.

The electrodes surface to system volume ratio: The effect of this ratio equal to 0.266 cm^2/L was investigated. As shown in the figure 13, in the absence of electrical current in the medium, the Fenton process occurs. The related process efficiency is very low equals to 22%. Then, after the electrical current generation, i.e., the occurrence of the Fered-Fenton process, the system efficiency increases noticeably and reaches 55%.

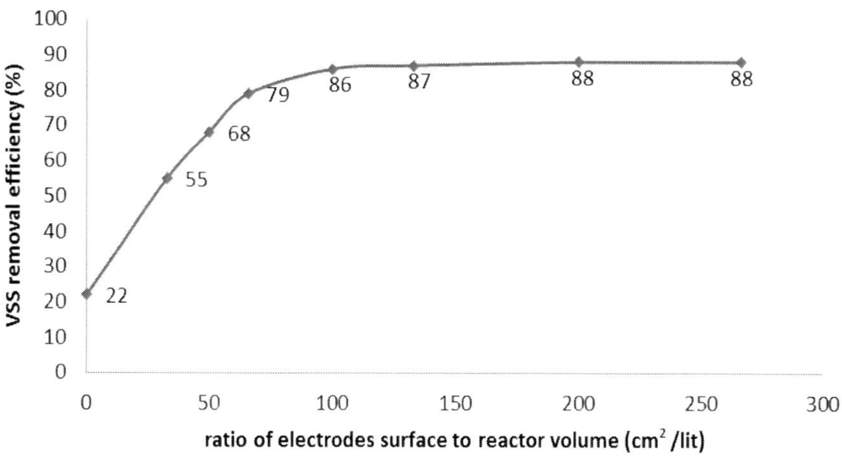

Figure 13. The relation between VSS removal efficiency and the ratio of electrodes surface to reactor volume (pH = 3.1, $[Fe^{2+}]/[H_2O_2] = 0.58$, current intensity = 650 mA, the retention time = 4 (hr), hydrogen peroxide concentration = 1568 mg/L, electrodes material: graphite, interelectrode distance = 1.5 cm, the number of Fenton's reagent injections = 5, sodium sulfate concentration = 0.11 mole/L).

As the electrodes surface increases in the range of 0-66 cm^2/L, the system efficiency steeply rises from 22% to 79% (57% improvement in the system efficiency). However, in the range of 66-100 cm^2/L a much gentler slope is noticed (7% rise in the system efficiency). Eventually, in the range of 100-200 cm^2/L the system efficiency rises only 2%. The figure 6 illustrates that increasing electrodes surface more than 200 cm^2/L has no effect on the system efficiency. Increasing electrodes surface is related inversely to the system resistance and directly to the current intensity (current density was considered constant, equals to 2.4 mA/ $cm^2_{electrode}$.L according to all experiments; therefore as the electrodes surface increases, the current intensity rises or vice versa), hence by increasing the electrodes surface, the system resistance decreases and the current intensity increases. Since power provided to the medium, according to the equation 15, it is related directly to the resistance and square of the current intensity. The effect of the current intensity increase overcomes the resistance decrease, subsequently power provided to the medium increases. Consequently, the concentration of the reduced ferric ion (the reaction 3) increases and the system efficiency rises respectively.

Changes in the rise of the system efficiency is considerable because in the low areas of the electrodes surface (0-66 cm^2/L), the rise in ferric ion concentration which is reduced due to the power increase in a certain retention time is very remarkable. However, as the electrodes surface increases from 66 to100 cm^2/L, the concentration's increase of the reduced ferric ion diminishes. In a wide range of electrodes surface, i.e from100 to 200cm^2/L, a high concentration of the present ferric ions is reduced and only a small amount of it has remained in the medium. Consequently, only a low percentage of the rise is observed. Due to the reduction of the ferric ion concentration at an electrodes surface equals to 200cm^2/L, the power increase does not affect the VSS removal efficiency. Therefore, the system efficiency does not increase after this level.

At the electrodes surfaces equal to 100 and 200cm^2/L, there is no significant difference in the system efficiency. Therefore, an electrodes surface equals to 100cm^2/L is selected for the reactor to lessen the electrodes consumption and consequently, the related cost.

The Trend of the Electrical Energy Consumption

The electrical energy consumption is one of the important issues according to the electrochemical systems. It needs to put more effort into improving the systems in this regard. Therefore, an analysis on the electrical energy consumption has been done as follows.

The solution resistance causes a potential drop which plays an effective role relating to the cell voltage (Daneshvar et al., 2007). Subsequently, the effect of the operation parameters including retention time, concentration of inflow organics, inter-electrode distance, sodium sulfate concentration and the ratio of electrodes surface to system volume on the electrical energy consumption is studied using the equation 23 (Ghosh et al., 2011).

$$UED = \frac{\frac{V \times I \times t}{v}}{\frac{VSS_{in}}{1000} \times \frac{\eta}{100}} \tag{23}$$

where the UED is the unit energy demand in kwh/kg VSS removal, V is the voltage in volt, *I* is the current intensity in Ampere, t is the retention time in hour, v is the reactor volume in Liter, VSS_{in} is the concentration of suspended solids in mg/L, and η is the VSS removal efficiency in percent.

It must be mentioned that except the concentration of the inflow organics, in other cases the trends of the electrical energy consumption for the removal of 1 kg VSS and the total electrical energy consumption are the same.

The retention time: The related experiments were conducted in 240 minutes. As shown in the Figure 7, the voltage decreases significantly in the first 90 minutes of the experiment; then from the 90th to 210th minute after the run, the voltage decrease becomes gentler and finally in the last 30 minutes becomes stable.

As the time passes, the system resistance decreases due to the reduction of organics concentration; therefore, potential drop decreases and the voltage of power supply decreases in the supply current intensity equals to 650 mA.

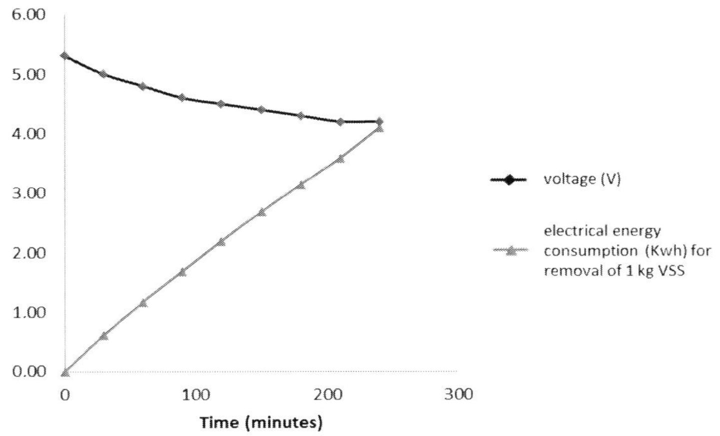

Figure 14. Alterations in the voltage and electrical energy consumption during 240 minutes (pH = 3.1, $[Fe^{2+}]/[H_2O_2] = 0.58$, the current intensity = 650 mA, the hydrogen peroxide concentration = 1568 mg/L, the electrodes material: graphite, the inter-electrode distance = 1.5 cm, the number of Fenton's reagent injection = 1, the ratio of electrodes surface to reactor volume = 266 cm^2/L).

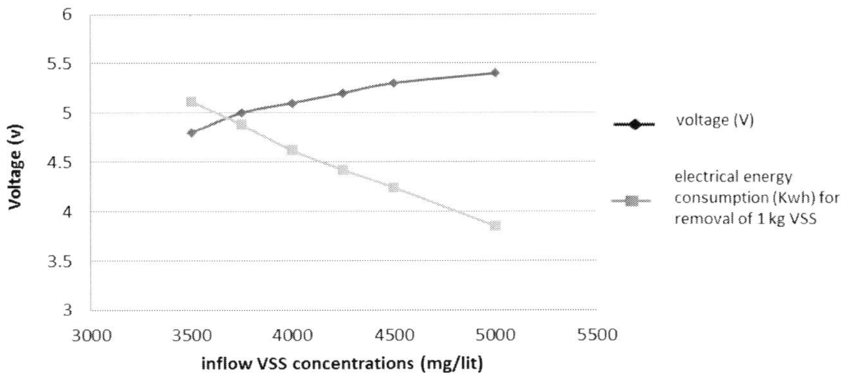

Figure 15. The relation between voltage, electrical energy consumption and inflow VSS concentration (pH = 3.1, $[Fe^{2+}]/[H_2O_2] = 0.58$, the current intensity = 650 mA, the retention time = 4 (hr), the hydrogen peroxide concentration = 1568 mg/L, the electrodes material: graphite, the interelectrode distance = 1.5 cm, the number of Fenton's reagent injection = 1, the ratio of electrodes surface to reactor volume = 266 cm^2/L).

Changes in the decreasing trend of the voltage declare that major concentration of the organics decreases in the first 90 minutes of the experiment. Regarding the reduction of the major organics concentrations in the first 90 minutes, the number of hydroxyl radicals collision with organic compounds decrease and the voltage decrease becomes gentler till after 210 minutes from the beginning of the experiment the organics concentration does not decrease anymore and consequently, voltage remains stable.

As the time passes, electrical energy consumption, as shown in the figure 14, increases since consumed electrical energy depends upon time as well as voltage (equation 23).

Concentrations of the inflow organics: The higher the concentration of inflow organics, the more the system resistance; consequently, the potential drop increases. Therefore, as illustrated in the Figure 15, the voltage of power supply increases for supplying constant current intensity equals to 650 mA.

However, since consumed electrical energy for removal of 1 kg VSS_{in} depends on the concentration of VSS_{in} and the removal efficiency as well as voltage, and considering that as the concentration of VSS_{in} increases, the VSS removal efficiency increases too (figure 8), the electrical energy consumption for the removal of 1 kg VSS decreases (while total electrical energy consumption increases).

The inter-electrode distance: The increasing distance between the electrodes causes slower movement of the produced ions and the system resistance increases which leads to the rising potential drop; hence the voltage of the power supply increases. Therefore, as shown in the figure 16, the consumed electrical energy based on the equation 23 increases.

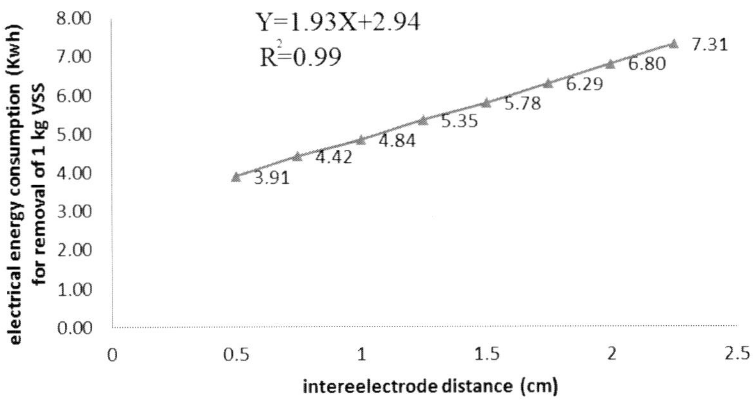

Figure 16. The relation between electrical energy consumption and interelectrode distance (pH = 3.1, $[Fe^{2+}]/[H_2O_2]$ = 0.58, the current intensity = 650 mA, the retention time = 4 (hr), the hydrogen peroxide concentration = 1568 mg/L, the electrodes material: graphite, the interelectrode distance = 1.5 cm, the number of Fenton's reagent injection = 1, the ratio of electrodes surface to reactor volume = 266 cm^2/L).

Electrical conductivity: By increasing sodium sulfate concentration, electrical conductivity of the solution increases, movement rate of produced ions accelerates, and the system resistance decreases which leads to a decreasing potential drop. Consequently, the voltage of the power supply decreases; the consumed electrical energy decreases also according to the equation 23 (figure 17).

The steep slope at the beginning of the figure 17 means that the initial addition of sodium sulfate which generates electrical conductivity causes a significant decrease of the electrical energy consumption (the consumed electrical energy decrease equals to 1.63 Kwh for each kg VSS removal). Due to the present electrical conductivity which already had been generated, the concentration of sodium sulfate increases more than 0.056 molar. However, the decreasing electrical energy consumption becomes very gentler (the consumed electrical energy decrease equals to 0.36 Kwh for each kg VSS removal). In other words, the generation of electrical conductivity greatly diminishes electrical energy consumption. However, its rise does not cause a noticeable decrease of the electrical energy consumption.

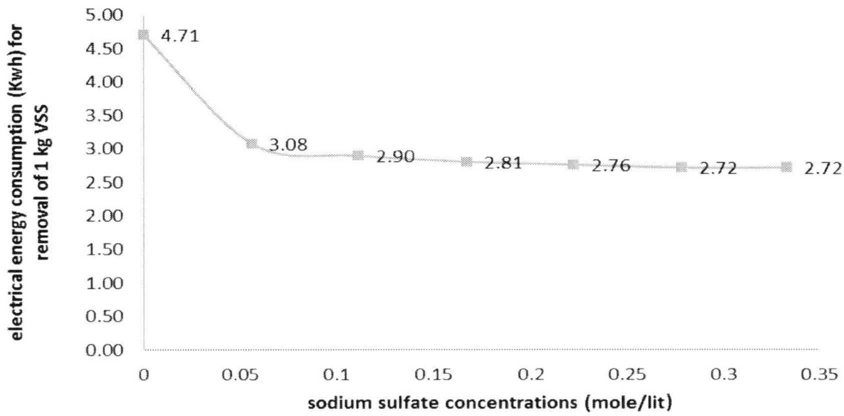

Figure 17. The relation between electrical energy consumption and sodium sulfate concentration (pH = 3.1, $[Fe^{2+}]/[H_2O_2] = 0.58$, the current intensity = 650 mA, the retention time = 4 (hr), the hydrogen peroxide concentration = 1568 mg/L, the electrodes material: graphite, the inter-electrode distance = 1.5 cm, the number of Fenton's reagent injection = 1, the ratio of electrodes surface to reactor volume = 266 cm^2/L).

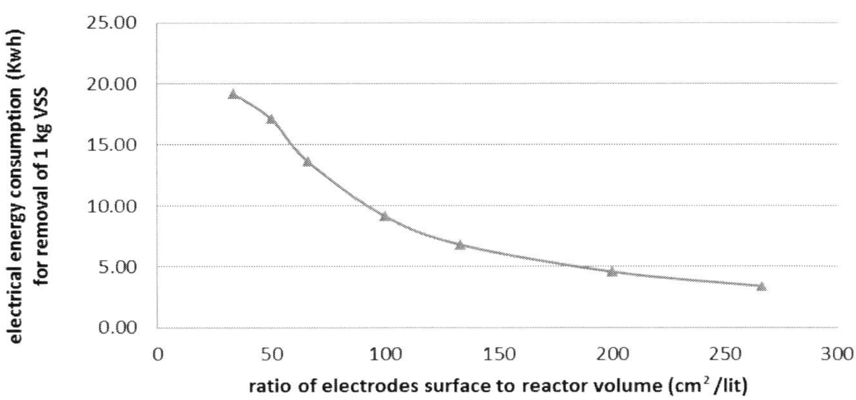

Figure 18. The relation between electrical energy consumption and the ratio of electrodes surface to reactor volume (pH=3.1, $Fe^{2+}/H_2O_2=0.58$, current intensity = 650 mA, retention time = 4 (hr), hydrogen peroxide concentration = 1568 mg/L, electrodes material: graphite, interelectrode distance = 1.5 cm, the number of Fenton's reagent injection = 1).

The area of electrodes surface: The increasing electrodes surface causes rising current intensity (in all experiments, the current intensity was considered constant, i.e., equals to2.4 $mA/cm^2_{electrode}.L$). Considering the constancy of the power supply, the voltage increases according to the equation 24. Therefore, according to the equation 23, the consumed electrical energy increases (figure 18).

$$R = \frac{V}{I} \tag{24}$$

3. Third Experimental Stage

During wastewater treatment, it is likely that many organic compounds are sorbed onto biological sludge. Consequently, they occur in much higher concentration than in the wastewater from which the sludge was derived. Organic matter in soil amendments can improve the properties and quality of soils, which is essential to guarantee long-term soil fertility (Draeger et al., 1999). In particular, an increase in organic matter content can improve physical properties (water retention, soil structure, water infiltration, bulk density, porosity), chemical properties (cation exchange capacity, pH) and, in some cases, biological properties (Moss et al., 2002). On the other hand, a high biological stability implies lower environmental impacts like odor generation, biogas production, leaching, and pathogen's re-growth during land application of the product (Moss et al., 2002). VSS removal efficiency has been experimentally studied in lab-scale system and, results to the optimum VSS removal equals to 81% (Badalians Gholikandi et al., 2014).

According to the increasing utilization of conventional activated sludge wastewater treatment technology worldwide, which leads to the generation of large quantities of waste activated sludge as an end-product of biological wastewater treatment processes, it is necessary to look for optimized methods in aim of the sludge volume reduction, materials and energy recovery, and the sludge reuse (Naoum et al., 2001).

Sustainable sludge management in an economically and environmentally effective way has become of increasing importance during the last decades (Cartmell et al., 2006). Sludge is composed of by-products accumulated at different stages of the wastewater treatment process. It contains both compounds of agricultural value (including organic matter, nitrogen, phosphorus and potassium, and to a lesser extent, calcium, sulphur, and magnesium), and various contaminants. Waste sludge is usually treated before disposal or reuse in order to reduce its water content, its fermentation propensity and the presence of pathogens. The sludge may undergo one or several treatment processes (US EPA 2001).

There is a great interest in spreading sludge on agricultural land due to the potential of recycling valuable components, i.e., organic matter, nitrogen, phosphorus and other plant nutrients (Fuentes et al., 2004). Concerning to the sludge land application, the phosphorus and nitrogen content is regarded as fertilizers that reduce the need for other types of fertilizers by 70% and 50%, respectively (Bengtsson et al., 1997).

According to the physical-chemical processes involved in activated sludge wastewater treatment method, trace elements like heavy metals that are existent in the wastewater tend to accumulate in the generated sludge. Therefore, repeated application of the sludge gradually increases the trace element content of the soil (Cheung 1988; Alvarez et al., 2002). Therefore, in order to retain the use of wastewater sludge in agricultural land, it is necessary to eliminate or minimize the amount of the contaminants, especially heavy metals in the sludge (Cheung 1988). In most cases of conventional sludge stabilization methods, there are some abstruse problems according to dewatering process, odor and pathogen elimination, organics and heavy metals removal units (Bureau et al., 2012). The problem is that the wastewater sludge constitutes a complex mixture that in addition to nitrogen and phosphorus often contains inorganic, i.e., metals and trace elements, and organic contaminants (Jones-Lepp and Stevens 2007). Many researchers have investigated the phytotoxic effect of heavy metals in soils amended with wastewater sludge (Lopez-Mosquera et al., 2000). Therefore, it is necessary to detect the trace elements concentration in end-product generated in sludge stabilization systems.

In order to manage the increasing amount of sludge product in hierarchy was proposed: (1) valorization in agriculture, (2) energetic valorization, and (3) landfilling. Energetic valorization is nowadays preferred to landfilling, since sludge recycling ensures the return of organic constituents, nutrients and micro elements to crop field that eases the substitution of chemical fertilizers (Union 2009). Finally, besides the environmental points of view, it should be considered that an acceptable sludge stabilization system has to provide an end-product suitable for land application with reasonable investment, operation and maintenance costs.

Sludge sampling, preparation, and characterization: Daily flow secondary sludge samples were taken from the effluent of the municipal wastewater treatment plant. Samples were analyzed in three different pH conditions including pH of 3, 5, and 7. According to previous research results, pH value equals to 3 is detected as optimized pH for operation of the system (Badalians Gholikandi et al., 2014). In addition, pH values equal to 5 and 7 were considered due to concerning with usual pH values of generated waste activated sludge in municipal wastewater treatment plants. After sludge sampling, the pH value of the samples were adjusted by adding an adequate dosage of Sulphuric acid and NaOH as required. Then fixed amounts of the Fenton reagents, i.e., $FeSO_4$ and hydrogen peroxide (H_2O_2), were added to the cell. Electrode plates were connected to laboratory model DC power supply apparatus and installed in cell. At last, voltage and current were adjusted that lead to system utilization at specified and optimized operational conditions (Badalians Gholikandi et al., 2014). Stabilized sludge samples were extracted from the cell after 6 hours, i.e., the optimized retention time, filtered using Whatman filter paper (grade 40).

In order to characterize the parameters, the samples were divided into two parts, i.e., the sludge cake (considered as biosolid) and the liquid. These two parts were determined after digestion using sulphuric acid (H_2SO_4). After the acidification, both liquid and sludge cake were analyzed in respect to the heavy metals content. The parameters of sludge quality were characterized in terms of pH, nitrate (NO_3^-),total nitrogen (TN), phosphate (PO_4^-), total phosphorus (TP), potassium (K), sodium (Na), calcium (Ca), heavy metals concentration and fecal bacteria indicators (fecal coliforms). The sludge samples were analyzed by analytical methods referred to the standard methods for the examination of water and wastewater (APHA, 1999; AWWA 2001). The concentrations of the heavy metals in all samples were determined by means of the Inductively Coupled Plasma (ICP) Model Varian-710, whereas the concentrations of the ions were measured by using Ion Chromatography Model Metrohm-IC850.

The experiments were divided into two parts. The first part was the study of the suitable pH value for the heavy metal removal and the second part was the detection of organic matter and nutrient concentrations in the investigated sludge samples before and after the operation of the Fered-Fenton system. Samples were subsequently equilibrated in distilled water for the pH and EC measurements. Fecal bacteria indicators were analyzed within 24 h after the sampling. The samples were stored at 4° C until analyzed. All experiments were performed at the temperature equals to 25°C and repeated three times in order to minimize possible errors. According to the samples volume and weight, the results are expressed on a dry matter basis (per kg or % dry substance of the sludge).

System performance at heavy metals removal: The main hazard associated with sludge application on agricultural land is the potential long term accumulation of toxic elements

(Singh and Agrawal, 2008), which may be up taken by plants and crops. Such elements include both inorganic contaminants such as heavy metals and organic microcontaminants. Currently, the heavy metals concentrations are regulated for land application of the wastewater sludge.

The bioavailability of heavy metals in soil and plants depends on soil pH, plant species and their cultivars, the growth stage, biosolids source, soil condition and the chemistry of the element (Warmar and Termeer, 2005). According to these authors, it is important to monitor the Cu and Zn concentrations in plant tissues after a few years of the sludge application to verify the tolerance levels for animal feed or human food. However, it is important to remark that the land sludge application must be done in accordance with the loading rates established by the regulations in force. Relating to the land application of biosolids, US EPA, Part 503 provides four sets of contaminants limits, i.e., contaminants ceiling concentration limits, contaminants concentration limits, cumulative contaminants loading rates, and annual contaminants loading rates listed in the table 3. All limits in the table are on a dry weight basis (US EPA 2001).

Table 3. Land application contaminants limites (US EPA 1993)

contaminant	ceiling concentration limits (mg/kg)	Contaminants concentration limites (mg/kg)	cumulative contaminants loading rates (mg/kg)	annual contaminants loading rates (mg/kg)
Arsenic	75	41	41	2
Cadmium	85	39	39	1.9
Chromium	3000	1200	3000	150
Copper	4300	1500	1500	75
Lead	840	300	399	15
Mercury	57	17	17	0.85
Molybdenum	75	-	-	-
Nickel	420	420	420	21
Selenium	100	36	100	5
Zink	7500	2800	2800	140

Table 4 shows the results of the heavy metals analyses of the influent sludge samples while the heavy metal concentrations of the final product are reported in the table 5.

Table 4. The heavy metal concentrations of the influent sludge samples

Heavy metals	Influent liquid (mg/kg)	Influent cake (mg/kg)
Copper	235	101
Lead*	100	86.1
Mercury*	3	-
Cobalt*	-	3
Molybdenum*	10	3
Chromium*	218	167
Zinc	750	213
Aluminum	49	17
Manganese	101	65

* Heavy metals are added to influent sludge samples synthetically.

Table 5. Heavy metal concentrations of the final treatment products samples

Heavy metals (mg/kg)	Effluent cake (pH=3)	Effluent liquid (pH=3)	Effluent cake (pH=5)	Effluent liquid (pH=5)	Effluent cake (pH=7)	Effluent liquid (pH=7)
Copper	59	43	72	25.5	80	9
Lead	19	56	31.7	18	-	-
Mercury	0.5	1	1	0.2	1	-
Cobalt	1.5	-	1	-	1	-
Molybdenum	3.2	7.5	4.1	3	5	2.1
Chromium	38	59	51.5	33	63	11
Zinc	96	303	113.5	178	147.5	121
Aluminum	11	27	23	13.5	30	5
Manganese	43	31	59	17	67	6

As expected, in all samples heavy metals concentrations were clearly below the law threshold; this is also probably attributed to the low concentration in the influent flow of sewage. According to the results, there were differences between influent sludge samples and the final treatment product, suggesting that heavy metals accumulation in sludge cake (which is considered as biosolid) is significant. Furthermore, the heavy metals concentrations in treated sludge samples are generally below the land application standards in this regard (Turovskiy and Mathai, 2006).

It is obvious that the rise in pH values leads to heavy metals concentration increase in the sludge cake, suggesting that the heavy metal precipitation has occurred. Hydroxide precipitation is the most common conventional treatment applied to the metal containing wastewaters for removing the heavy metals. This method is based on low solubility of metal hydroxides at alkaline pH values. Therefore, it can be derived from the results that Fered-Fenton process cause a greater heavy metal accumulation in sludge cake at pH value equals to 7, due to the sulfide precipitation conducted under high pH levels. It has also been established that the pH has a considerable influence on the performance of electrocoagulation and reduction process (Dermentzis et al., 2011).

The results are in accordance with previous studies (Adhoum 2004). Major part of the heavy metals removal such as Cu^{2+} is contributed to electro-deposition. This process is a clean and effective technology due to its recovery of heavy metals through the redox reactions without the disadvantages of conventional chemical precipitation processes.

The system efficiency on nutrients removal: Wastewater sludge provides essential nutrients for the plant growth. Biosolids are able to restore nitrogen, phosphorus, sulphur and other nutrients in soils. The concentration of nutrients in the sludge depends on wastewater composition and the treatment process, and on subsequent sludge management. Nitrogen, phosphorus, potassium, calcium, and sodium concentrations resulting from this study are shown in the table 6.

Table 6. Nutrients usual amounts of the generated sludge in municipal wastewater plants (Bina et al., 2005)

Parameter	Carbon	Nitrogen	Phosphorus	Sodium	Potassium	C/N
Amount (%)	8 - 50	0.1 - 3	0.35 – 3.5	0.2 – 0.5	0.1 – 2.8	20

In general, the sludge is characterized by a considerable variability in nutrients content, depending on the wastewater source and treatment processes (Moss et al., 2002). The concentration of nutrients is needed to ensure appropriate dosages of the sludge for agricultural application. It should be remarked that the required agricultural doses are frequently dependent on the fertilizer and soil characteristics (Pomares and Canet 2001).

Although they are essential for the plant growth; nutrients (particularly nitrogen and phosphorus) can be harmful when excessively applied. Different studies have shown both nitrogen and phosphorous accumulation in the sludge amended soils (Hernández et al., 1999). It is well known that the over-application of nitrogen can lead to nitrate contamination of groundwater, although such a risk is reduced if nutrients are applied at agronomic rates (Moss et al., 2002).

The sludge organics content of different parts (sludge liquid and cake) of the influent samples are shown in table 7. As mentioned, the sludge samples were analyzed in three different pH values, i.e., 3, 5, and 7. The results related to the system efficiency on sludge organics content are expressed in the tables 8, 9, and 10 respectively.

Table 7. Organic matter content in different parts of samples

sample	Nitrate(%TS)	TN(%TS)	Phosphate(%TS)	TP(%TS)
Influent liquid	0.02	1.16	0.3	0.96
Influent cake	0.33	3.95	0.68	2.01
Total	0.35	5.11	0.98	2.97

Table 8. Organic matter content in different parts of the samples (pH=3)

sample	Nitrate(%TS)	TN(%TS)	Phosphate(%TS)	TP(%TS)
effluent liquid	0.15	2.01	0.16	0.91
effluent cake	0.03	0.26	0.33	0.98
Total	0.18	2.27	0.49	1.9

Table 9. Organic matter content in different parts of the samples (pH=5)

sample	Nitrate(%TS)	TN(%TS)	Phosphate(%TS)	TP(%TS)
effluent liquid	0.1	1.73	0.11	0.48
effluent cake	0.08	0.97	0.49	1.49
Total	0.18	2.7	0.6	2

Table 10. Organic matter content in different parts of the samples (pH=7)

sample	Nitrate(%TS)	TN(%TS)	Phosphate(%TS)	TP(%TS)
effluent liquid	0.05	1.54	0	0.15
effluent cake	0.19	2.01	0.65	1.89
Total	0.24	3.55	0.65	2.04

Nitrogen comes from microbial biomass present in the sludge and from wastewater residues. It is obvious from the results that increasing the pH lead to nitrate and total nitrogen reduction in the liquid, while these concentrations have a reasonable increase in sludge cake suggesting that organic matter content accumulation happened probably due to sludge mineralization and ammonification. However, compared to nitrate system removal efficiency at high pH, treatment process is more effective at low pH on removal percentage of nitrate concentrations.

Phosphorus in sludge comes from biomass formed during wastewater treatment, residues, and phosphate-containing detergents and soaps. Biosolids can be considered as a phosphorus source assuring a slow and continued release to plants and crops (Andreoli et al., 2007). In this study, the total phosphorus (TP) values showed no significant differences between the influent samples and the treated sludge samples. Therefore, according to the results, it should be noticed that the system effect on phosphate and total phosphorus concentrations is negligible.

The system efficiency on pathogens removal: Since the treated sludge may contain high numbers of pathogens, depending on the treatment processes employed, limited values for fecal bacteria indicators have also been suggested. The pathogens associated with municipal wastewater sludge present a potential problem at amended soils (Pahren et al., 1979). The US EPA, Part 503 requires the reduction of potential disease-bearing microorganisms called pathogens (such as bacteria and viruses) in sewage sludge. In this regard, pathogens reduction requirements for the wastewater sludge are divided into two categories, i.e., Class A and Class B. When the related goals achieved, Class A wastewater sludge can be land applied without any pathogen-related restrictions on the site (US EPA 1995). The implicit goal of the Class B requirements is to ensure that pathogens have been reduced to a level that is unlikely to pose a threat to public health and the environment under specific use conditions (US EPA 1995).

The part 503 establishes alternatives for demonstrating that wastewater sludge meets Class A pathogen reduction requirements. The rule requires that the density of fecal coliforms must be less than 1,000 Most Probable Number (MPN) per gram (g) total solids (dry weight), and it must be satisfied immediately after the treatment process is completed (US EPA 1995).

Compliance with the fecal coliform limitation for Class B, biosolids must be demonstrated by calculating the geometric mean of samples. The geometric mean must be less than 2,000,000 MPN or colony-forming units (CFU) per g of dry substances (DS). The Table 11 expresses the pathogen content, i.e., the fecal coliforms of the sludge samples before and after the treatment process occurrence.

Table 11. The sludge samples number of fecal coliforms

sample	pH	fecal coliforms (MPN/g.DS)
Influent sludge	3	7032
Influent sludge	7	7032
Treated sludge	3	837
Treated sludge	7	2056

It is obvious from the table 11 that the sludge samples treated in pH values equals to 3, meets the Class A pathogens reduction requirements while the sludge samples treated in pH values equals to 7 meets the Class B pathogens reduction requirements. Therefore, the results obtained from this study ensure that the wastewater sludge pathogens have been reduced to the levels that are unlikely to cause a threat to public health and the environment.

REFERENCES

Ai, Z., Mei, T., Liu, J., Li, J., Jia, F., Zhang, L. & Qiu, J. (2007) Fe@ Fe_2O_3 core-shell nanowires as an iron reagent. 3. Their combination with CNTs as an effective oxygen-fed gas diffusion electrode in a neutral electro-Fenton system, *The Journal of Physical Chemistry C* 111(40), American Chemical Society (ACS), 14799-14803.

Alvarez, E.A., Mochón, M.C., Sanchez, J.C. & Rodrıguez, M.T. (2002) Heavy metal extractable forms in sludge from wastewater treatment plants, *Chemosphere* 47(7), Elsevier, 765-775.

Andreoli, C.V., Pegorini, E.S., Fernandes, F. & Santos, H.F. (2007) Land application of sewage sludge, IWA Publishing, London, UK.

Anotai, J., Su, C.C., Tsai, Y.C. & Lu, M.C. (2010) Effect of hydrogen peroxide on aniline oxidation by electro-Fenton and fluidized-bed Fenton processes, *Journal of Hazardous Materials* 183(1), Elsevier, 888-893.

APHA (1999) Standard Methods for the Examination of Water and Wastewater, 20th edition, American Public Health Association /Water Environment Federation, Washington DC, USA.

Atkins, P. (1997) Physical Chemistry, 6th edition,W.H. *Freeman and Company*, New York, USA.

Babuponnusami, A., Muthukumar, K. (2012) Advanced oxidation of phenol: a comparison between Fenton, electro-Fenton, sono-electro-Fenton and photo-electro-Fenton processes, *Chemical Engineering Journal* 183, Elsevier, 1-9.

Badalians Gholikandi, G., Masihi, H., Azimipour, M., Abrishami, A. & Mirabi, M. (2014) Optimizing stabilization of waste-activated sludge using Fered-Fenton process and artificial neural network modeling (KSOFM, MLP), *Environmental Science and Pollution Research*, 21(11), Springer, 7177-7186.

Bengtsson, M., Lundin, M. & Molander, S. (1997) Life Cycle Assessment of Wastewater Systems: Case Studies of Conventional Treatment, Urine Sorting and Liquid Composting in Three Swedish Municipalities,Chalmers University of Technology, Gothenburg, Sweden.

Bina, B., Movahedian Atar, H. & Amini, A. (2005) Investigation the quality of dried activated sludge in Isfahan WWTPs, *Water and Wastewater journal*, 49.

Brillas, E., Sirés, I., & Oturan, M. A. (2009) Electro-Fenton process and related electrochemical technologies based on Fenton's reaction chemistry, *Chemical Reviews* 109(12), American Chemical Society (ACS), 6570-6631.

Bureau, M.A., Drogui, P., Sellamuthu, B., Blais, J.F. & Mercier, G. (2012) Municipal wastewater sludge stabilization and treatment using electrochemical oxidation technique,

Journal of Environmental Engineering 138(7), American Society of Civil Engineers (ASCE), 743-751.

Carrère, H., Dumas, C., Battimelli, A., Batstone, D.J., Delgenès, J. P., Steyer, J.P. & Ferrer, I. (2010) Pretreatment methods to improve sludge anaerobic degradability: a review, *Journal of Hazardous Materials* 183(1), Elsevier, 1-15.

Cartmell, E., Gostelow, P., Riddell-Black, D., Simms, N., Oakey, J., Morris, J. & Pollard, S.J. (2006) Biosolids a fuel or a waste? An integrated appraisal of five co-combustion scenarios with policy analysis, Environmental Science & Technology 40(3), *American Chemical Society* (ACS), 649-658.

Cheng-chun, J. and Jia-fa, Z. (2007) Progress and prospect in electro-Fenton process for wastewater treatment. *J. Zhejiang Univ. Sci. A* 8(7):1118–1125.

Cheung, Y.H. (1988) Acid treatment of anaerobically digested sludge: effect on heavy metal content and dewaterability, *Environment International* 14(6), Elsevier, 553-561.

Daneshvar, N., Khataee, A.R., Amani Ghadim, A.R., & Rasoulifard, M.H. (2007) Decolorization of CI Acid Yellow 23 solution by electrocoagulation process: Investigation of operational parameters and evaluation of specific electrical energy consumption (SEEC), Journal *of Hazardous Materials* 148(3), Elsevier, 566-572.

Dermentzis, K., Valsamidou, E., Lazaridou, A. & Kokkinos, N. (2011) Nickel removal from wastewater by electrocoagulation with aluminum electrodes, *Journal of Engineering Science and Technology Review* (JESTR) 4(2),188-192.

Draeger, K.J., Pundsack, J.W., Jorgenson, M.D. & Mulloy, P.J. (1999) Watershed effects of biosolids land application: literature review, Water Environment Research Foundation.

Eberson, L. (1987) *Electron Transfer Reactions in Organic Chemistry,* Springer, Berlin, Germany.

Fuentes, A., Lloréns, M., Sáez, J., Soler, A., Aguilar, M.I., Ortuño, J.F. & Meseguer, V.F. (2004) Simple and sequential extractions of heavy metals from different sewage sludges, *Chemosphere* 54(8), Elsevier, 1039-1047.

Ghoneim, M.M., El-Desoky, H.S. & Zidan, N.M. (2011) Electro-Fenton oxidation of Sunset Yellow FCF azo-dye in aqueous solutions, *Desalination* 274(1), Elsevier, 22-30.

Ghosh, P., Samanta, A.N. & Ray, S. (2011) Reduction of COD and removal of Zn^{2+} from rayon industry wastewater by combined electro-Fenton treatment and chemical precipitation, *Desalination* 266(1), Elsevier, 213-217.

Gu, L., Zhu, N., Guo, H., Huang, S., Lou, Z. & Yuan, H. (2013) Adsorption and Fenton-like degradation of naphthalene dye intermediate on sewage sludge derived porous carbon, *Journal of Hazardous Materials* 246, Elsevier, 145-153.

He, M.H., Wei, C.H. (2010) Performance of membrane bioreactor (MBR) system with sludge Fenton oxidation process for minimization of excess sludge production, *Journal of Hazardous Materials* 176(1), Elsevier, 597-601.

Hernandez, T., Moreno, J.I. & Costa, F. (1999) Influence of sewage sludge application on crop yields and heavy metals availability, *Journal of soil science and plant nutrition* 37 (2), BIOREN-UFRO, 201–210.

Jones-Lepp, T.L., Stevens, R. (2007) Pharmaceuticals and personal care products in biosolids/sewage sludge: the interface between analytical chemistry and regulation, *Analytical and Bioanalytical Chemistry* 387, Springer, 1173–1183.

Li, X., Xing, M., Yang, J. & Huang, Z. (2011) Compositional and functional features of humic acid-like fractions from vermicomposting of sewage sludge and cow dung, *Journal of Hazardous Materials* 185(2), Elsevier, 740-748.

Liu, W., Ai, Z., & Zhang, L. (2012) Design of a neutral three-dimensional electro-Fenton system with foam nickel as particle electrodes for wastewater treatment, *Journal of Hazardous Materials* 243, Elsevier, 257-264.

Lopez-Mosquera, M.E., Moiron, C. & Carral, E. (2000) Use of dairyindustry sludge as fertilizer for grasslands in northwest Spain: heavy metal level in the soil and plant, *Resources, Conservation and Recycling* 30, Elsevier, 95–109.

Modirshahla, N., Behnajady, M. & Mohammadi-Aghdam. S. (2008) Investigation of the effect of different electrodes and their connections on the removal efficiency of 4-nitrophenol from aqueous solution by electrocoagulation, *Journal of Hazardous Materials154*, Elsevier, 778-786.

Moss, L.H., Epstein, E. & Logan, T. (2002) Evaluating Risks and Benefits of Soil Amendments Used in Agriculture, *International Water Association (IWA) and Water Environmental Research Foundation (WERF),* Alexandria.

Muruganandham, M., Swaminathan, M. (2004). Decolourisation of Reactive Orange 4 by Fenton and photo-Fenton oxidation technology, *Dyes and Pigments* 63(3), 315-321.

Naoum, C., Fatta, D., Haralambous, K.J. & Loizidou, M. (2001) Removal of heavy metals from sewage sludge by acid treatment, *Journal of Environmental Science and Health,* Part A 36(5), 873-881.

Narayanan, T. S., Magesh, G. & Rajendran, N. (2003) Degradation of O-chlorophenol from aqueous solution by electro-Fenton process, Fresenius *Environmental Bulletin* 12(7), 776-780.

Nidheesh, P.V., Gandhimathi, R. (2012) Trends in electro-Fenton process for water and wastewater treatment: an overview, *Desalination* 299, Elsevier, 1-15.

Pahren, H.R., Lucas, J.B., Ryan, J.A. & Dotson, G.K. (1979) Health Risks Associated with Land Application of Municipal Sludge, *Journal of the Water Pollution Control Federatmn* 51, 2588-2601.

Pomares, F., Canet, R. (2001) Organic waste utilization in agriculture: origin, composition and characterization (In Spanish), Lleida, Spain: In: Boixadera, J., Teira, M.R. (Eds.).

Rabaaoui, N., Saad, M.E.K., Moussaoui, Y., Allagui, M. S., Bedoui, A. & Elaloui, E. (2013) Anodic oxidation of o-nitrophenol on BDD electrode: Variable effects and mechanisms of degradation, *Journal of Hazardous Materials* 250, Elsevier, 447-453.

Samaras, P., Papadimitriou, C.A., Haritou, I. & Zouboulis, A.I. (2008) Investigation of sewage sludge stabilization potential by the addition of fly ash and lime, *Journal of Hazardous Materials* 154(1), Elsevier, 1052-1059.

Shemer, H., Linden, K.G. (2006) Degradation and by-product formation of diazinon in water during UV and UV/H_2O_2 treatment, *Journal of hazardous materials* 136(3), 553-559.

Shen, J., Ou, C., Zhou, Z., Chen, J., Fang, K., Sun, X. & Wang, L. (2013) Pretreatment of 2, 4-dinitroanisole (DNAN) producing wastewater using a combined zero-valent iron (ZVI) reduction and Fenton oxidation process, *Journal of Hazardous Materials* 260, 993-1000.

Singh, R. P., Agrawal, M. (2008) Potential benefits and risks of land application of sewage sludge, *Waste management* 28(2), Elsevier, 347-358.

Stock, J.T., Orna, M.V. (1989) Electrochemistry, past and present 390, Washington DC, *American Chemical Society* (ACS).

Su, C.C., Chang, A.T., Bellotindos, L.M., & Lu, M.C. (2012) Degradation of acetaminophen by Fenton and electro-Fenton processes in aerator reactor, *Separation and Purification Technology* 99, 8-13.

Tao, H.C., Wei, X.Y., Zhang, L.J., Lei, T. & Xu, N. (2013) Degradation of p-nitrophenol in a BES-Fenton system based on limonite, *Journal of Hazardous Materials* 254, 236-241.

Tokumura, M., Katoh, H., Katoh, T., Znad, H. T. & Kawase, Y. (2009) Solubilization of excess sludge in activated sludge process using the solar photo-Fenton reaction, *Journal of Hazardous Materials* 162(2), 1390-1396.

Turovskiy, I.S., Mathai, P.K. (2006) *Wastewater Sludge Processing*, Wiley-Interscience.

U.S. EPA (1993) *Standards for the Use or Disposal of Sewage Sludge*, Federal register 40 CFR Part 503, 58 FR9248 to 9404.

U.S. EPA (2001) disposal and recycling routes for sewage sludge, part 3, scientific and technical report. Luxembourg, office for official publications of European communities.

U.S. EPA. (1995). *Process Design Manual for Sludge Treatment and Disposal*, EPA 625/1-75/003a.

Union, C.O. (2009) Green Paper on the management of bio-waste in the European Union, 2953rd Environment Council meeting.

Vanysek, P. (2011) Electrochemical Series in Handbook of Chemistry and Physics, 92nd Edition, Chemical Rubber Company.

Wahl, A. (2005) A short history of electrochemistry, *Galvanotechtnik* 96, 1820-1828.

Wang, C.T., Chou, W.L., Chung, M.H., & Kuo, Y.M. (2010) COD removal from real dyeing wastewater by electro-Fenton technology using an activated carbon fiber cathode, *Desalination* 253(1), Elsevier, 129-134.

Wang, C.T., Hu, J.L., Chou, W.L., & Kuo, Y.M. (2008) Removal of color from real dyeing wastewater by Electro-Fenton technology using a three-dimensional graphite cathode, *Journal of Hazardous Materials* 152(2), Elsevier, 601-606.

Wang, H., Wang, C., Chen, F., Ma, M., Lin, Z., Wang, W., ... & Wang, X. (2012) Modification to degradation of hexazinone in forest soils amended with sewage sludge, *Journal of Hazardous Materials*, Elsevier 199, 96-104.

Wang, L., Zheng, Z., Zhang, Y., Chao, J., Gao, Y., Luo, X. & Zhang, J. (2013) Biostabilization enhancement of heavy metals during the vermiremediation of sewage sludge with passivant, *Journal of Hazardous Materials* 244, Elsevier, 1-9.

Warman, P.R., Termeer, W.C. (2005) Evaluation of sewage sludge, septic waste and sludge compost applications to corn and forage: Ca, Mg, S, Fe, Mn, Cu, Zn and B content of crops and soils, *Bioresource technology* 96(9), 1029-1038.

Weng, H.X., Ma, X.W., Fu, F.X., Zhang, J.J., Liu, Z., Tian, L.X. & Liu, C. (2014) Transformation of heavy metal speciation during sludge drying: Mechanistic insights, *Journal of Hazardous Materials* 265, Elsevier, 96-103.

Xiao, G., Ni, M.J., Chi, Y., Jin, B.S., Xiao, R., Zhong, Z.P., & Huang, Y.J. (2009) Gasification characteristics of MSW and an ANN prediction model, *Waste Management* 29(1), Elsevier, 240-244.

Xu, G., Liu, M., & Li, G. (2013) Stabilization of heavy metals in lightweight aggregate made from sewage sludge and river sediment, *Journal of Hazardous Materials* 260, Elsevier, 74-81.

Xu, J.Q., Yu, R.L., Dong, X.Y., Hu, G.R., Shang, X.S., Wang, Q., & Li, H.W. (2012) Effects of municipal sewage sludge stabilized by fly ash on the growth of Manilagrass and transfer of heavy metals, *Journal of Hazardous Materials* 217, Elsevier, 58-66.

Yu, Y., Chan, W.I., Liao, P.H., & Lo, K.V. (2010) Disinfection and solubilization of sewage sludge using the microwave enhanced advanced oxidation process, *Journal of Hazardous Materials* 181(1), Elsevier, 1143-1147.

Zhang, H., Zhang, D., & Zhou, J. (2006) Removal of COD from landfill leachate by electro-Fenton method, *Journal of Hazardous Materials* 135(1), Elsevier, 106-111.

Zhou, M., Yu, Q., Lei, L., & Barton, G. (2007) Electro-Fenton method for the removal of methyl red in an efficient electrochemical system, *Separation and Purification Technology* 57(2), 380-387.

In: Enhanced Electrochemical Advanced Oxidation ... ISBN: 978-1-63482-089-9
Editor: Gagik Badalians Gholikandi

Chapter 5

SLUDGE STABILIZATION AND REUSE EMPLOYING FERED-FENTON TECHNOLOGY: AN INVESTIGATIVE VIEW TO APPLICATIONS OF ARTIFICIAL NEURAL NETWORKS MODELING AND TAGUCHI METHOD

Gagik Badalians Gholikandi**, *Mohammad Azimipour* and *Hamidreza Masihi
Faculty of Water Engineering and Environment,
Shahid Beheshti University, A.C., Tehran, Iran

INTRODUCTION

Modeling process possesses various definitions in different areas. It can be defined as the process of the fundamental knowledge application or the experience to simulate or describe the performance of a real system to achieve certain goals (Khandan 2010); it can be also defined as the creation of the set of relationships that relate inputs to the process and its outputs. Such a relationship can be constructed either analytically or empirically. For the former, the relationship is based on analysis of the interactions between the input variables, and for the latter it is based on the basis of experimental/historical data (Ramsey 1994).

The practical use of the simulation modeling is closely associated with the development of the computer technology, as the development of computers and simulation programs was one of the main reasons for the related methodology progress and its practical applications improvement (Domonkos 2010).

Modeling as a cost-effective and significant tool has been a main component of the real systems organizing measurements for understanding the related causes and effects. Modeling is the art of approximating a system for the purpose of learning its characteristics (Ramsey 1994; Atkinson et al., 2004). The main feature of a model for the real-life situations is its applicability for solving problems of industry-relevant sizes (Schichl 2004).

* Corresponding author: Hakimieh- Tehranpars, Shahid Abbaspour Blvd., PO Code 16589-53571, PO Box 16765-1719, Tehran, Iran. Email: g.badalians@yahoo.com.

Over the last two decades, the significant development in information technology in different areas, especially the control of the environmental processes and the rapid development of the hardware made research links between environmental and computer scientists possible (Cortes et al., 2000). One of the new disciplines is the environmental informatics (Radermacher et al., 1994) that combines various research fields, such as artificial intelligence (AI), geographical information systems (GIS), modeling and simulation, user interface, etc. One of the most practical modeling approaches is the artificial intelligence. Its applications to environmental modeling have increased due to its capabilities (Walker 1993). A wide range of the artificial intelligence techniques, such as case-based reasoning, rule-based systems, artificial neural networks, genetic algorithms, cellular automata, fuzzy models, Bayesian networks and data mining relating to environmental sciences and engineering can be found in several works (Chen et al., 2008; Jakeman et al., 2008).

In this chapter, two approaches to modeling are applied, i.e., the artificial neural network (ANN) as an empirical modeling method and the Taguchi method as a statistical modeling technique. Various applications of these methods for wastewater treatment, especially sludge treatment are considered accordingly.

Artificial Neural Network (ANN)

In 1943, a simple neural network model with electronic circuits was proposed (McCulloch and Pitts 1990). It was the beginning of the networks development such as Adaptive Linear Element (ADALINE) by Widrow and Perceptron by Rosenblat (Rosenblatt 1958; Widrow 1962). After a 10-year lag, proposing new models by other researchers as Grossberg, Kohonen and Hopfiled had started a new era of the neural networks (Haykin 2004). There are three main paradigms for the artificial intelligence systems, i.e., symbolic approaches, connectionist approaches and the combination of the mentioned approaches (Hanrahan 2011). One of the connectionist approaches is the neural network. It is inspired by the biology of human brain (Jahandideh et al., 2009). A neural network consists of simple processing elements called neurons which are arranged in layers (Khataee and Kasiri 2010). Each neuron is connected to the neurons of another layer with varying coefficients called weights that store the knowledge of a neural network (Holubar et al., 2002). The artificial neural networks are classified in different categories based on various characteristics. However, the main feature is the learning algorithm by which there are supervised and unsupervised (Badalians Gholikandi et al., 2014a). The most widely applied artificial neural network model for the supervised category is the Multi-Layer Perceptron (MLP). Due to the capabilities of the MLP, it is a popular choice for the environmental engineering systems modeling (Zeng et al., 2003; Azimipour et al., 2014). This feed forward neural network consists of the input layer, the output layer and one or more hidden layers. While the number of input and output layer neurons are dependent on nature of the problem, previous researches have shown that one hidden layer is sufficient for the environmental issues. Moreover, due to the universal approximation theory, a network with a single hidden layer and a sufficiently large number of neurons can be used for any input-output structure interpretation (Salari et al., 2005).

The most widely used supervised learning algorithm which is considered the workhorse of the ANNs is Back Propagation (BP) (Rumelhart et al., 1986; Badalians Gholikandi et al., 2011; Badalians Gholikandi and Khosravi 2012; Tashaouie et al., 2012). However, because of some drawbacks, new algorithms have been created and applied successfully. One of these algorithms is Levenberg-Marquardt (L-M). This kind of the algorithm is a combination of the steepest descent method and the Gauss–Newton algorithm and possesses the speed advantage of the Gauss–Newton algorithm and the stability of the steepest descent method. This method can be applied from 10 to 100 times more time-effective than usual gradient descent back propagation method (Pendashteh et al., 2011).

Another specific training algorithm is unsupervised learning algorithm. The most related applied network is the Kohonen Self-Organizing Feature Maps (KSOFM). Willshaw-von der Malsburg is another kind of unsupervised networks. The KSOFM neural network technique consists of two layers, i.e., the input layer and the competitive layer of neurons (Ghaseminezhad and Karami 2011). It is usually a two-dimensional grid. The array of the neurons called the topographic map attempts to preserve the neighborhood relationships between the input and output spaces. This principle is called feature maps (Hong et al., 2003). The ability of learning and organizing information without applying the corresponding output indicates the self-organizing term (Grieu et al., 2006; Ghaseminezhad and Karami 2011). The lattice of the neurons can be illustrated as rectangular or hexagonal but hexagonal organization is used most often due to the better presentation of connections between neighboring neurons (Heikkinen et al., 2011). For each neuron (n), there is an n-dimensional weight $W_i = [W_{n1}, W_{n2},\ldots,W_{nm}]^T$ where m is the dimension of the input vectors (Hong et al., 2003). In the output layer, categorizations (grouping) of the inputs are formed (Dayhoff, 1990) and therefore, outputs are clusters or categories into which the input falls with another depending on its similarity (Hong et al., 2003). By this kind of the algorithm, regularities or trends in the input signals are being looked for. For the performance of the unsupervised learning, a competitive learning rule must be utilized in which the best matching unit (BMU) can be chosen with "winner-takes-all" strategy (Haykin 2004; Rustum et al., 2008; Llorens et al., 2008). Two different training algorithms for the KSOFM neural network are the sequential training and the batch training. They are the different versions of the SOM algorithm (Vesanto et al., 1999; Garcia and Gonzalez 2004).

For both supervised and unsupervised neural networks there are two parameters for evaluating the performance of the models. The following two statistical indices are used for the MLP neural network performance, i.e., Mean Square Error (MSE, equation 1) and R (Regression Coefficient, equation 2) respectively.

$$MSE = \frac{1}{N}\sum_{i=1}^{N}(T_i - O_i)^2 \qquad (1)$$

$$R = \frac{\sum_{i}^{n}(T_i - \overline{T}_i)(O_i - \overline{O}_i)}{\sqrt{\sum_{i}^{n}(T_i - \overline{T}_i)^2 \sum_{i}(O_i - \overline{O}_i)^2}} \qquad (2)$$

T and O indicate the experimental and network outputs, $\overline{T}$ and $\overline{O}$ are the average of the mentioned data, respectively. N is the total number of the data.

The following two error criteria are normally used regarding the KSOFM neural network method, i.e., the quantization error and the topographical error as defined by the equations 3 and 4 respectively (Garcia and Gonzalez 2004, Rustum et al., 2008).

$$q_e = \frac{1}{N}\sum_{i=1}^{N} \| x_i - m_c \| \quad (3)$$

$$t_e = \frac{1}{N}\sum_{i=1}^{N} u(x_i) \quad (4)$$

q_e= quantization error,
N = number of samples,
x_i = i-th data sample or vector,
m_c = prototype vector of the best matching unit for xi,
||…|| denotes distance,
t_e = topographical error,
u = binary integer, which is equal to 1 if the first and the second BMU for the argument of u are not adjacent unit on the map, otherwise i = 0.

TAGUCHI METHOD

Taguchi experimental design is a simple method established on optimization of design, quality and cost (Zirehpour et al., 2014). Taguchi method emphasizes on quality design during the process designing. Taguchi results include optimal conditions of VSS removal, the effect of each parameter on system output and the response estimation in optimal conditions. The tool used by Taguchi for analysis of experimental results is signal- to-noise (S/N) analysis method which is the ratio of operating fixed factors to uncontrollable chaos factors. Taguchi method also uses another powerful tool, the ANOVA analysis of variance which is applied for analyzing the results (Roy 1990).

The most important step in the design of the experiments is the selection of the control factors, i.e., [Fe]/[H_2O_2], pH, [H_2O_2] (mg/L) and current density (mA). Each of these parameters was defined at four levels which are listed in the table 1. Accordingly, the orthogonal L_{16} array was selected for design of the experiments (table 2). While the total numbers of tests without using Taguchi design method are (4^4) = 256 states, by using it, the number of tests has dropped to 16 cases. In this study, the experiments were performed with three times repeat that in total, 48 (16 × 3) tests were performed. According to preliminary investigations and analyses performed by Taguchi method, the retention time parameter has little effect on removal efficiency; thus, it was not considered as an effective parameter. In all experiments performed, sludge retention time of the reactor was considered constant (the retention time equals to 240 min.).

Table 1. Controllable factors and their levels

Number	Factors	Level 1	Level 2	Level 3	Level 4
1	$[Fe]/[H_2O_2]$	0.145	0.435	0.58	1.45
2	pH	1	3	5	9
3	$[H_2O_2]$	313	1255	1568	2196
4	Current density	250	650	850	1250

Table 2. Experimental layouts of an L_{16} orthogonal array

Trial	$[Fe]/[H_2O_2]$	pH	$[H_2O_2]$	Current density
1	1	1	1	1
2	1	2	2	2
3	1	3	3	3
4	1	4	4	4
5	2	1	2	3
6	2	2	1	4
7	2	3	4	1
8	2	4	3	2
9	3	1	3	4
10	3	2	4	3
11	3	3	1	2
12	3	4	2	1
13	4	1	4	2
14	4	2	3	1
15	4	3	2	4
16	4	4	1	3

For the reason that any VSS removal increment will increase the system performance, higher signal-to-noise (S/N) ratio is better. The S/N ratio was calculated based on the following equation 5:

$$\frac{S}{N} = 10Log_{10}\left(\frac{1}{n}\sum \frac{1}{Y_i^2}\right) \tag{5}$$

where n is the number of repetitions under the same experimental conditions, and y represents each of measured results.

For the recognizing optimum condition, analysis of mean (ANOM) was utilized (Zolfaghari et al., 2011). Calculation of the mean of S/N which is required for each factor at all levels is based on equation 6 (Taguchi 1990):

$$(M)_{Factor=I}^{Level=i} = \frac{1}{n_{Ii}} \sum_{j=1}^{n_{Ii}} \left[\left(\frac{S}{N}\right)_{Factor=I}^{Level=i} \right]_j \tag{6}$$

M is the mean of the S/N ratio of factor I in the level of i.

For the statistical survey, analysis of variance (ANOVA) is also used. Based on the equation7, the influence of each factor (parameter) on VSS removal is calculated.

$$p_F = \frac{SS_F - (DOF_F V_{Er})}{SS_T} \times 100 \quad (7)$$

P_F indicates percentage contribution of each factor and DOF_F represents the degree of the freedom for each factor which is obtained by the equation 8:

$$DOF_F = (L-1) \quad (8)$$

L is the number of the levels for each factor (For all factors, L=4).

The factorial sum of squares SS_F which is used in the equation 7 is given by the equation 9:

$$SS_F = \frac{mn}{L}\sum_{k=1}^{L}(\overline{Y}_K^F - \overline{Y}_T)^2 \quad (9)$$

$\overline{Y}_K^F$ is the average value of the measurement results of a certain factor in the k_{th} level. The numbers of the experiments are represented by m and the numbers of the repetitions are indicated by n. $\overline{Y}_T$ is obtained by equation10:

$$\overline{Y}_T = \frac{1}{mn}\sum_{j=1}^{m}\left(\sum_{i=1}^{n} Y_i\right) \quad (10)$$

The parameter SS_T in the equation 7 is calculated by the equation11 as follows:

$$SS_T = \sum_{j=1}^{m}\left(\sum_{i=1}^{n} Y_i^2\right)_j - mn(\overline{Y}_T)^2 \quad (11)$$

The variance of error which in represented by V_{Er} in the equation 7 is given by the equation 12:

$$V_{Er} = \frac{SS_T - \sum_{F=A}^{D} SS_F}{m(n-1)} \quad (12)$$

The variance of each factor, V_F is also given by the following equation 13:

$$V_F = \frac{SS_F}{DOF_F} \tag{13}$$

Variance ratio (F_F) is calculated by the equation 14:

$$F_F = \frac{V_F}{V_{Er}} \tag{14}$$

If calculated F_F (equation 14) is less than extracted F from statistical Handbook in certain confidence level, this factor has no effect on dependent variable (VSS); otherwise VSS is affected by F (Roy 1990).

APPLICATIONS OF THE MODELING RELATING TO WASTEWATER TREATMENT AND WASTE SLUDGE STABILIZATION

In the last decade, various modeling approaches, e.g., MLP, KSOFM, Fuzzy logic, RSM, Taguchi, and its combination were employed regarding wastewater treatment and sludge stabilization processes. Some of the related publications are listed in the following tables 3, 4, 5, and 6.

Table 3. MLP modeling approaches

Modeling Approach	Modeling Target	Reference
MLP	Predicting and improving the operating performance of a WWTP	(Hamed et al., 2004; Vyas et al., 2011)
	Describing the nonlinearity of complex wastewater treatment processes of a WWTP	(Choi and Park 2001)
	Acquiring the knowledge base of a real WWTP	(Mjalli et al., 2007)
	Providing a tool to form a basis for controlling the operation of the process	(Nasr et al., 2007)
	Applying for simulation-based WWTP design and simulation of integrated urban wastewater systems	(Raduly et al., 2007)
	Evaluating the performance of each treatment plant units and estimating qualitative profile of wastewater	(Mehrdadi et al., 2012; Hasanlou et al., 2012)
	Representing the behavior of the treatment system	(Teze et al., 2010)
	Providing a simple and practical way for the operation and management in wastewater treatment plant	(Zhang and Hu 2012)
	Useful tool for accurate and cost-effective modeling of biochemical processes	(Lee et al., 2002)

Table 3. (Continued)

Modeling Approach	Modeling Target	Reference
MLP	Developing four independent multivariate models that are able to predict the dynamics of BOD, COD, SS, and TN removal in a WWTP	(Lee et al., 2011)
	Predicting the environmental properties of the output stream from the WWTP	(Gontarski et al., 2000)
	Simulating and optimizing of biogas production process from the digester	(Abu Qdais et al., 2010)
	Modeling and controlling methane production in anaerobic digesters	(Holubar et al., 2002)
	Predicting of hydrogen sulfide and ammonia in biogas from anaerobic digestion	(Dellana et al., 2009)
	Predicting the methane fraction in landfill gas originating from field-scale landfill bioreactors	(Ozkaya et al., 2007)
	Implementing a model to simulate the dynamic behavior of a large-scale anaerobic sewage sludge digestion system	(Güclü et al., 2011)
	Modeling and estimating biogas production rate in a thermophilic upflow anaerobic sludge blanket digester	(Kanat and Saral 2009)
	Simulating and optimizing design and configuration of anaerobic baffled reactor (ABR)	(Badalians Gholikandi et al., 2014b)
	Providing better predictions of nitrogen contents in treated effluents to be used for groundwater recharge	(Chen et al., 2003)
	Controlling UV and UV-TiO_2 disinfection processes to meet three total coliform count limits for three wastewater reuse purposes	(Lin et al., 2012)
	Modeling sludge bulking of activated sludge	(Capodaglio et al., 1991)
	Describe the sedimentation process and providing a control of minimum amount of reagents	(Syu et al., 2003)
	Describing ammonia emissions generated in composting sewage sludge	(Boniecki et al., 2012)
	Predicting the sludge volume index	(Lou and Zhao 2012)
	Controlling the Fenton process for textile wastewater treatment	(Yu et al., 2010)
	Modeling of the behavior of complex systems and predicting of advanced oxidation process of organics in water matrix	(Bolanča et al., 2014)
	Modeling and Optimization of Heterogeneous Photo-Fenton Process	(Kasiri et al., 2008)
	Modeling treatment of recalcitrant wastewater by advanced oxidation processes	(Elmolla et al., 2010)

Modeling Approach	Modeling Target	Reference
MLP	Approximation of the reduction of cod effluents from the manufacturing of polyesters through photo-Fenton process/ Ozonization	(Silva et al., 2011)
	Modeling for decoloration by UV/H_2O_2	
	Optimization of the AZO dyes decoloration process by determining of the H_2O_2 addition critical point	
	Discoloration process modeling	
	Modeling of C.I. Direct Red 16 (DR16) decolorization using Fentonic reagents in a batch reactor	(Saien et al., 2012)
	Modeling of water and wastewater treatment using various homogeneous and heterogeneous nanocatalytic processes	(Khataee and Kasiri 2010)
	Modeling the kinetics of a photochemical water treatment process	(Göb et al., 1999)
	Prediction and simulating oil degradation in aqueous solution by photo-Fenton process	(Mustafa et al., 2014)
	Predicting the methyl tert-butyl ether (MTBE) concentration during irradiation time in optimized conditions	(Salari et al., 2005)

Table 4. KSOFM and combined modeling approaches

Modeling Approach	Modeling Target	Reference
KSOFM + MLP	Estimating the process efficiency and providing reliable information for the plant monitoring	(Grieu et al., 2006)
Mechanistic + MLP	Simulating the performance of a WWTP with substantial influent fluctuations	(Fang et al., 2011)
MLP + ASM1	Modeling of activated sludge process (ASP)	(Moral et al., 2008)
Self-Organizing Fuzzy Neural Networks (SOFNN)	Improving the control efficiency of Activated Sludge wastewater treatment	(Qiao and Wang 2008)
KSOFM	Analyzing an activated sludge treatment process	(Heikkinen et al., 2011)
KSOFM	Estimating nitrogen and organic matter concentrations in the stream and to describe the river water quality	(Llorens et al., 2008)
KSOFM	Development of software sensors for the rapid prediction of BOD	(Rustum et al., 2008)
KSOFM	Analysing the multidimensional process data, and diagnosing the inter-relationship of the process variables in a real activated sludge WWTP	(Hong et al., 2003)
KSOFM	Classifying operational data of WWTP and determining the reasons for high effluent concentrations of BOD, TSS and fecal coliform	(Cinar 2005)

Table 4. (Continued)

Modeling Approach	Modeling Target	Reference
KSOFM	Elucidating heavy metal removal mechanisms and predicting heavy metal concentrations in experimental constructed wetlands	(Lee and Scholz 2006)
KSOFM	Assessing the efficiency of chlorinated solvent removal from petrochemical wastewater in a refinery WWTP	(Tobiszewski et al., 2012)
KSOFM	Development of plant supervision techniques for implementation in a WWTP	(Garcia and Gonzalez 2004)
KSOFM + MLP	Determining the correlations of the parameters and evaluating investigating parameters correlation and prediction aims towards proper improving of excess biological sludge stabilization	(Badalians Gholikandi et al., 2014a)

Table 5. Taguchi method

Modeling Approach	Modeling Target	Reference
Taguchi	Planning a minimum number of experiments and optimizing desalination by vacuum membrane distillation	(Mohammadi and Safavi 2009)
	Optimizing oil extraction from oil sludge with biosurfactant formulas	(Zheng et al., 2012)
	Optimizing olive oil mill wastewater treatment and determining the most significant parameters	(Zirehpour et al., 2014)
	Determining optimal conditions Photo-Fenton and Fenton processes for dairy effluent treatment	(Loures et al., 2013)
	Determining optimum conditions Fenton process in removing total petroleum hydrocarbons (TPH) in oily sludge	(Farzadkia et al., 2014)
	studying anaerobic process evaluation and process optimization	(Venkata et al., 2005)
	Reviewing applications in various biotechnology segments : fermentation, food processing, molecular biology, wastewater treatment and bioremediation processing, molecular biology, wastewater treatment and bioremediation	(Rao et al., 2008)
	Determining optimum conditions for methylene blue dye removal from aqueous solutions by electrocoagulation	(Asghari et al., 2012)
	Featuring the application in the effluent treatment of polyester resin that is originated from textile industries and through the application of Advanced Oxidation Processes	(de Freitas et al., 2013a)
	Demonstrating the application in effluent treatment of polyester resin by using advanced oxidation processes (heterogeneous photocatalysis – UV/TiO_2)	(de Freitas et al., 2013b)

Modeling Approach	Modeling Target	Reference
Taguchi	Obtaining the optimum conditions of coagulation–flocculation process in tannery wastewater treatment	(Aber et al., 2010)
	Optimizing of wastewater treatment using spiral-wound reverse osmosis element	(Madaeni and Koocheki 2006)
	Determining optimum conditions for color removal from textile dye bath house effluents in a zeolite fixed bed reactor	(Engin et al., 2008)
	Finding optimal mixing proportion of combined treatment of domestic wastewater with landfill leachate and the optimal operating conditions for the removal efficiencies of nitrogen and evaluating the influence of parameters in an anaerobic–anoxic–aerobic (A^2/O) bioreactor system	(Yu et al., 2010)

Table 6. RSM and RSM-ANN methods

Modeling Approach	Modeling Target	Reference
RSM	Optimizing electrochemical treatment conditions for the treatment of water-based paint wastewater	(Körbahti et al., 2007)
	Optimizing sludge conditioning with optimized combinations of a polymer and skeleton builders prior to filter pressing	(Benítez et al., 1994)
	Optimizing coagulation–flocculation process and evaluating the effects and interactions of operational factors on the treatment efficiency	(Wang et al., 2007)
	Fenton's preoxidation for the oil mill wastewater	(Ahmadi et al., 2005)
	Optimizing electrochemical treatment of textile dye wastewater	(Körbahti 2007; Körbahti and Tanyolaç 2008)
	Optimizing the operating variables in coagulation–flocculation treatment of stabilized leachate	(Ghafari et al., 2009)
	Evaluating and optimizing the interactive effects of operating variables on physical and oxidative performances of Fenton process for treatment of landfill leachate	(Wu et al., 2010)
	Finding optimal operating point for maximum biodegradability of pulp and paper industry wastewater	(Soloman et al., 2009)
	Predicting the behavior of an upflow anaerobic sludge blanket (UASB) bioreactor for the treatment of petroleum refinery effluent	(Rastegar et al., 2011)
	Modeling and analyzing anaerobic digestion of pretreated Palm oil mill effluent (POME)	(Zinatizadeh et al., 2006)

Table 6. (Continued)

Modeling Approach	Modeling Target	Reference
RSM	Optimization of growth conditions for bacterial strains for enhancement of aerobic digestion potential of waste activated sludge interceded through Ethylene diamine tetra acetic acid (EDTA)	(Kavitha et al., 2013)
	Determining the optimum conditions for the Acid Black 172 dye removal efficiency from aqueous solution by electrocoagulation	(Taheri et al., 2012)
	Optimization of Fenton and electro-Fenton oxidation of biologically treated coking wastewater	(Zhu et al., 2011)
RSM + ANN	Evaluating and comparing the modeling efficiency of ANN and RSM for modeling efficiencies for biohydrogen production on Sugar Cane Molasses	(Whiteman and Kana 2014)
	Modeling electrocoagulation system for the removal of copper from synthetic wastewater	(Bhatti et al., 2011)
	Optimizing methane gas production from Co-Digestion of food waste and poultry manure	(Yusof et al., 2014)
	Developing an approach for the evaluation of fluoride adsorption process	(Mourabet et al., 2014)
	Modeling and optimizing the photocatalytic degradation of total phenolic (TPh) compounds in real secondary and tertiary treated municipal wastewaters	(Antonopoulou et al., 2012)

Results of the Fered-Fenton Experiments Modeling

Artificial Neural Networks Modeling MLP

In order to determine the optimized structure of the network, trial-and-error approach was employed including different constructing networks by varying the number of hidden neurons and the activation functions (Jahandideh et al., 2009). For this purpose, three types of the activation functions were used. For the hidden layer 'tansig' and 'logsig' were utilized and for the output layer 'purelin' was used besides the two aforementioned transfer functions. The numbers of the hidden layer neurons were varied from 6 to 20, while the number of input and output neurons depends on the nature of the problem (Salari et al., 2005).

In this study, the development of the ANN model for determining the best fitted model is carried out as follows (Delnavaz et al., 2010; Badalians Gholikandi et al., 2011): (1) Input selection and preprocessing in the range of [-1, 1], (2) Creating different models (networks) by varying the number of the hidden neurons and the activation functions, (3) Model validation by means of the early stopping technique, (4) Evaluating the performance of the models by R^2 and MSE.

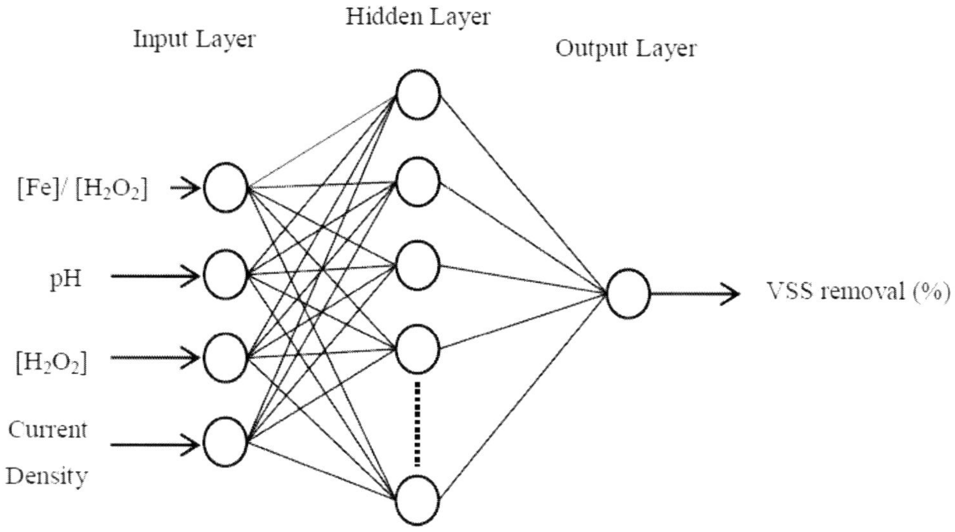

Figure 1. Structural organization (Topological Architecture) of Neural Network model.

In present study ANN-based model trained and tested using MATLAB®, was fed with parameters mentioned in table 1, [Fe]/ [H_2O_2] ((mg/L)/(mg/L)), [H_2O_2] (mg/L), pH, and I (mA). Therefore, there are 48 data sets for the simulation which were normally divided into three groups, i.e., training, validation and test subsets to overcome over-fitting issue. In this research 60%, 15% and 25% of data were used for training, validation and testing, respectively. The best model architecture which has the lowest MSE (MSE = 2.64) was obtained of 4:14:1 structure and is shown in Figure 1. The activation function of this model is 'tansig' at hidden layer and 'purelin' at output layer. The R^2 value of 0.99 for training data and 0.98 for validation and test subsets show high performance of chosen model. Figure 2 shows correlations between predicted and measured (experimental) VSS removal percentage for data subsets ((a) Test, (b) Validation and (c) All data).

Relative Importance: For the determination of the relative importance of input variables on VSS removal percentage, the Garson equation 15 was applied which uses the weights of the best trained network (Elmolla et al., 2010).

$$I_j = \frac{\sum_{m=1}^{m=Nh}((\left|W_{jm}^{ih}\right|\sum_{k=1}^{Ni}\left|W_{km}^{ih}\right|)\times\left|W_{mn}^{ho}\right|)}{\sum_{K=1}^{k=Ni}[\sum_{m=1}^{m=Nh}((\left|W_{km}^{ih}\right|/\sum_{k=1}^{Ni}\left|W_{km}^{ih}\right|)\times\left|W_{mn}^{ho}\right|)]} \tag{15}$$

I_j is the relative importance of the j_{th} input variable on the output variable, N_i and N_h are the number of input and hidden neurons and the W is for connection weights, respectively. The superscripts $'i'$, $'h'$ and $'o'$ are for input, hidden and output layers, respectively and subscripts $'k'$, $'m'$ and $'n'$ refer to input, hidden and output neurons, respectively (Delnavaz et al., 2010). Table 7 shows the best ANN model weights which were used in the equation 15.

Table 7. The weights between Input & Hidden Layers (W1) and between Hidden & Output Layers (W2)

Neuron	W1					W2	
	Input Variables					Output	
	[Fe]/[H_2O_2]	pH	[H_2O_2]	I	Bias	Neuron	VSS removal (%)
1	0.3355	-0.4989	-2.7912	0.0573	-2.5568	1	-0.4013
2	-1.2664	0.3997	-1.9419	1.4856	2.2235	2	0.3887
3	1.0174	1.6333	-1.8846	-0.5698	-1.811	3	0.0849
4	0.2644	1.4698	-1.5173	1.2101	-1.7626	4	0.2607
5	-1.1072	0.1185	-2.3767	0.6186	1.0304	5	-0.2897
6	2.3947	-1.1623	-0.4259	-0.2714	-0.6178	6	-0.1213
7	2.0165	1.2976	-1.2938	-0.1773	-0.2656	7	-0.456
8	1.6044	-0.4606	0.0767	2.1433	0.3127	8	-0.6626
9	-1.4291	0.1349	1.4733	-1.7377	-0.6064	9	-0.299
10	-1.9105	1.6991	-1.041	-0.7809	-0.5495	10	-0.8028
11	1.6846	2.2367	-0.7745	-0.5571	0.899	11	0.3126
12	0.9841	2.025	1.3879	0.4513	1.8594	12	-0.2574
13	0.8746	1.6421	-0.7032	1.7217	2.3934	13	0.2336
14	0.0184	1.751	-1.9971	-0.1338	2.7303	14	-0.5119
						Bias	-0.0775

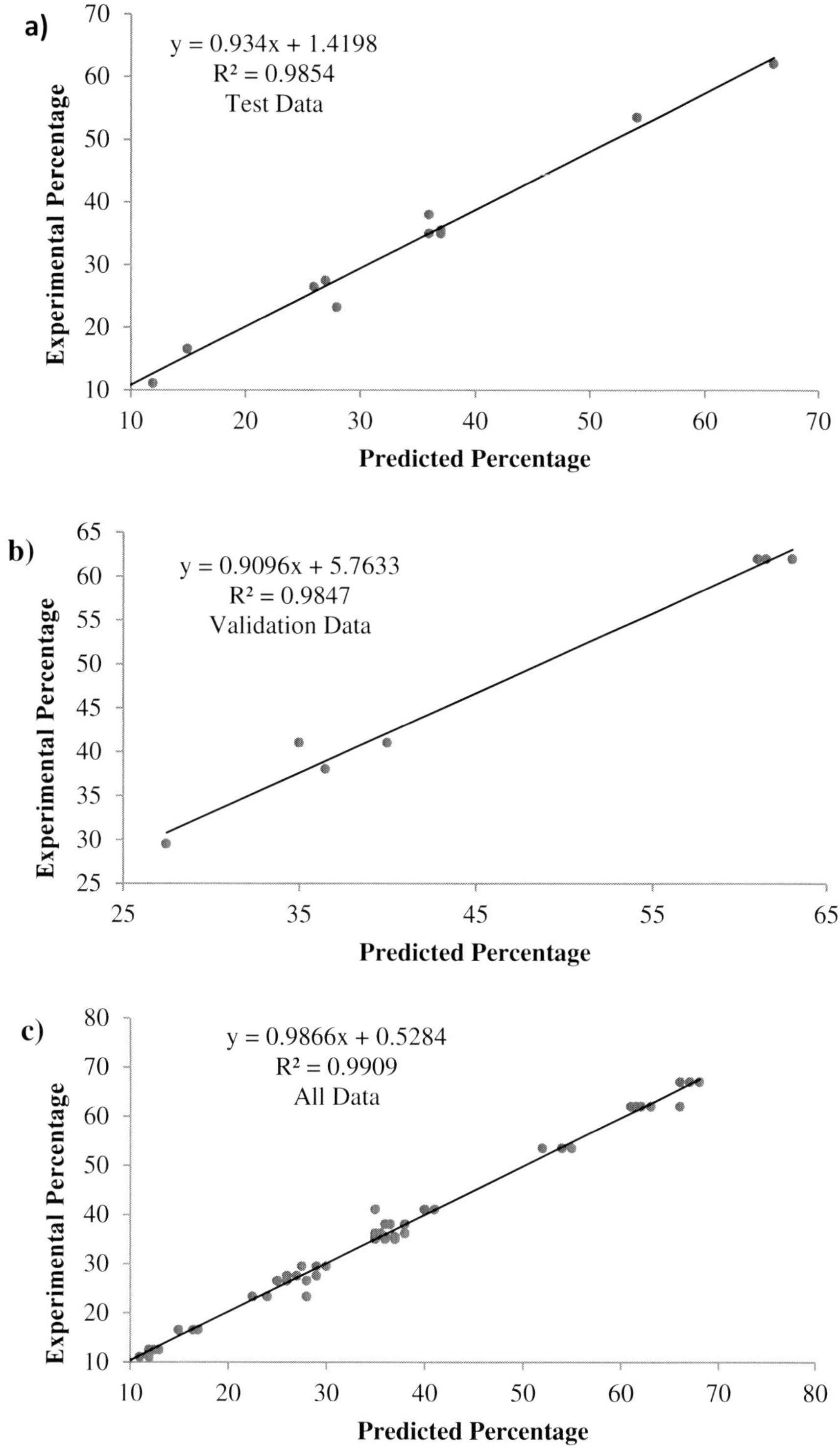

Figure 2. The ANN Regression analysis: (a) test (b) validation and (c) all data.

Table 8. The relative importance of input variables

Input variables	Importance (%)
$[Fe]/[H_2O_2]$	25.93
pH	24.43
$[H_2O_2]$	30.38
I	19.25
Total	100

The table 8 shows the influence of each input variable.

In this chapter, for more investigation and evaluation of VSS removal, MLP neural network has been utilized in which two parameters were changed simultaneously and three other parameters were considered constant. Therefore, these figures are applicable in operation of the reactor in order to prediction VSS removal efficiency in various conditions.

The figure 3 shows the increasing system removal efficiency depending on the rising retention time. Likewise the removal efficiency decreases depends on increasing $[Fe^{2+}]/[H_2O_2]$. As shown in figure 4 both parameters $[Fe^{2+}]/[H_2O_2]$ and $[H_2O_2]$ have the most influence in a specific range (table 4). Figure 5 shows that both parameters $[Fe^{2+}]/[H_2O_2]$ and current density have the most influence in a specific range too (table 4). Figure 6 shows the VSS removal efficiency increases related to increasing retention time and the best performance has been achieved in a specific range of H_2O_2 concentration (table 4). Figure 7 shows that both parameters, i.e., H_2O_2 concentration and current density reached to their optimum in the particular range of VSS removal efficiency.

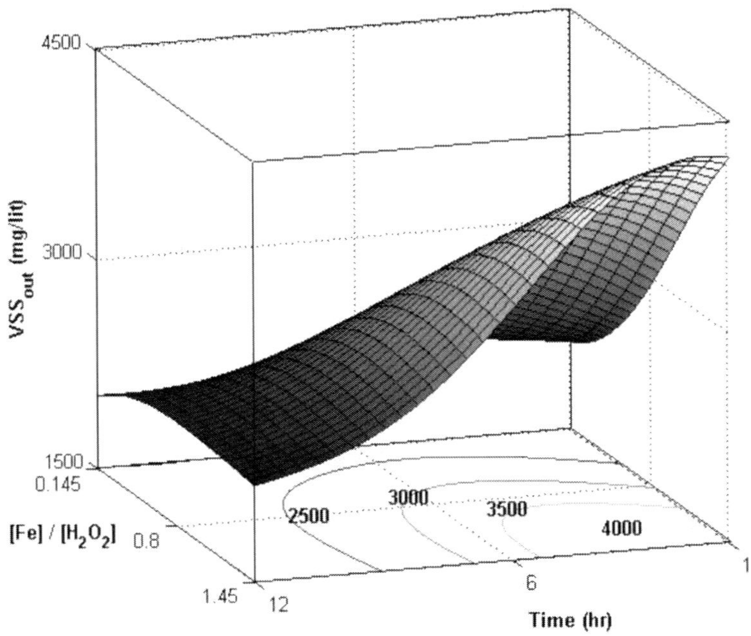

Figure 3. The simultaneous effect of $[Fe^{+}]/[H_2O_2]$ and Time on VSS_{out}.

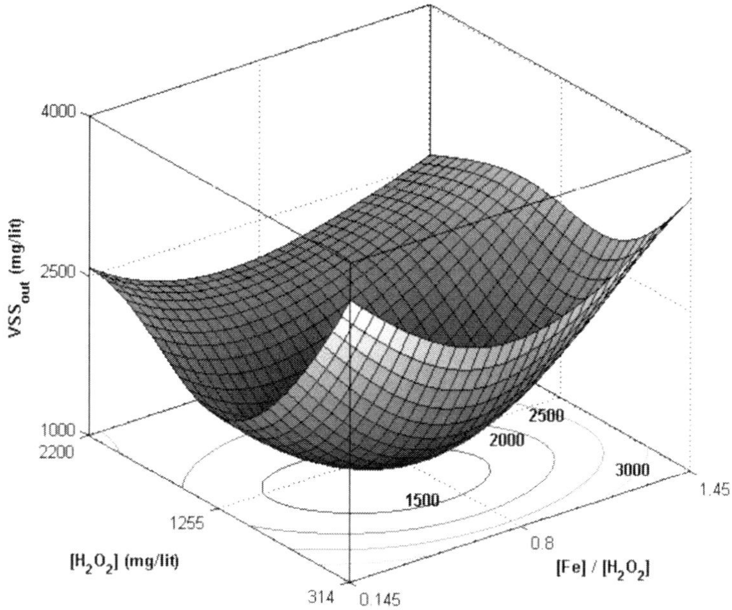

Figure 4. The simultaneous effect of $[Fe^{+}]/[H_2O_2]$ and $[H_2O_2]$ on VSS_{out}.

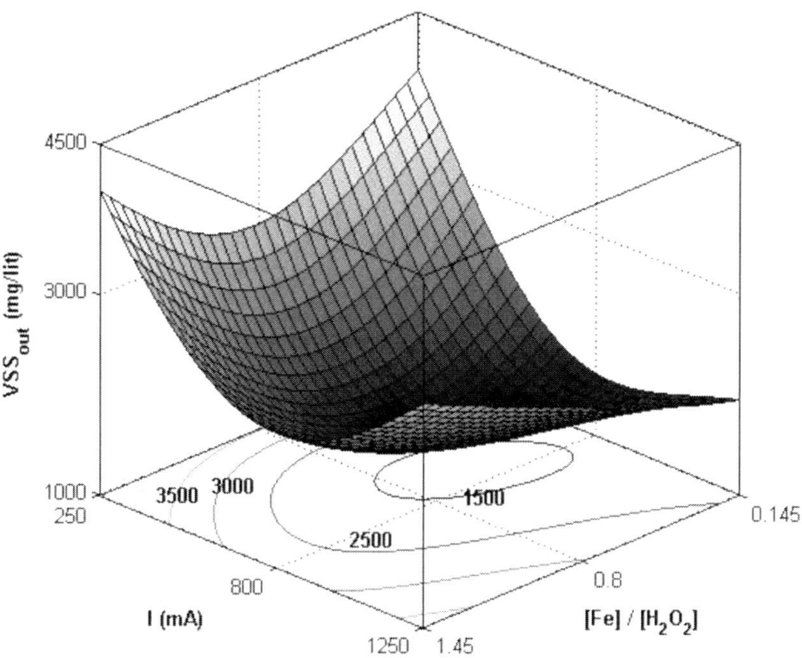

Figure 5. The simultaneous effect of $[Fe^{+}]/[H_2O_2]$ and current density on VSS_{out}.

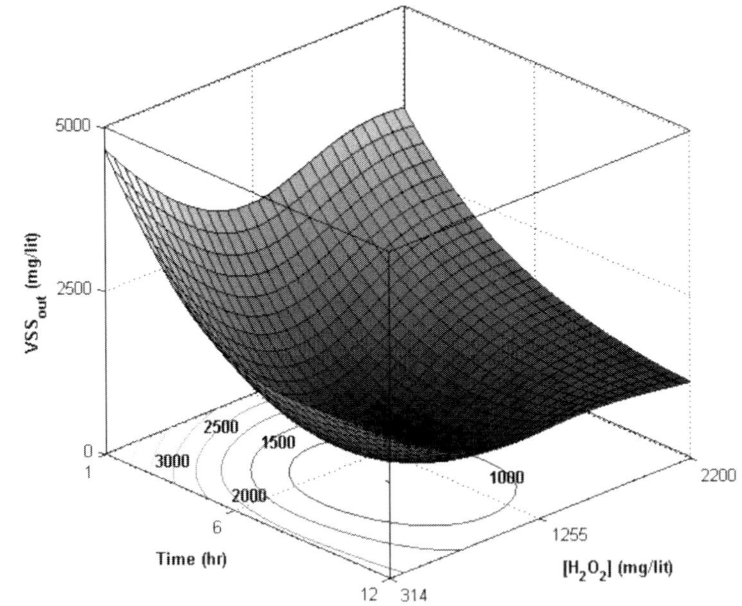

Figure 6. The simultaneous effect of Time and $[H_2O_2]$ on VSS_{out}.

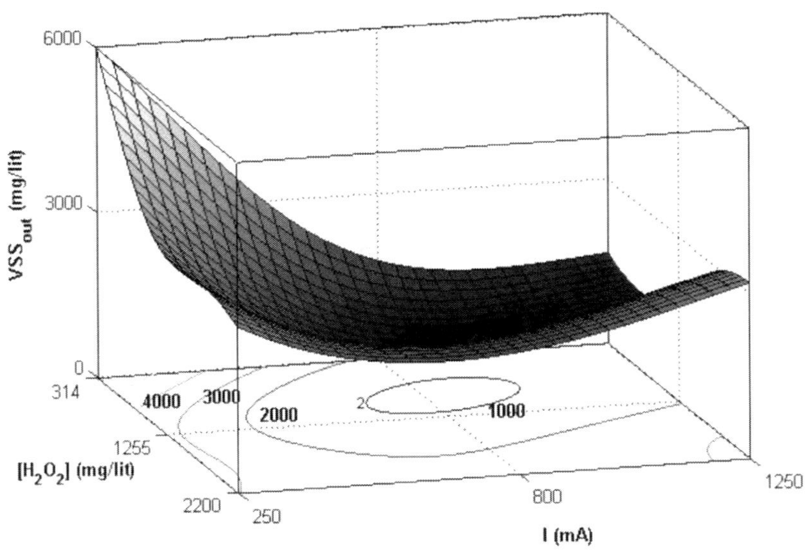

Figure 7. The simultaneous effect of $[H_2O_2]$and Current Density on VSS_{out}.

In the table 9, the optimum range has been presented in order to attain the maximum VSS removal efficiency based on neural network modeling. The mentioned results demonstrate the high flexibility of the system in its high range of the VSS removal.

Table 9. Final Results of the ANN modeling

Constant Parameters			Variable Parameter		Optimum Range	Minimum VSS (out)
I	$[H_2O_2]$	Time	$[Fe^{2+}]/[H_2O_2]$	pH	$1 < pH < 4.3$ $0.63 < [Fe^{2+}]/[H_2O_2] < 0.93$	$1450 < VSS_{out} < 1750$
I	$[H_2O_2]$	$[Fe^{2+}]/[H_2O_2]$	Time	pH	$1 < pH < 3$ $8 < Time < 10$	$600 < VSS_{out} < 1000$
I	Time	$[Fe^{2+}]/[H_2O_2]$	$[H_2O_2]$	pH	$4 < pH < 6$ $1555 < [H_2O_2] < 1950$	$1250 < VSS_{out} < 1500$
$[H_2O_2]$	Time	$[Fe^{2+}]/[H_2O_2]$	I	pH	$3.4 < pH < 6.3$ $700 < I < 900$	$1400 < VSS_{out} < 1600$
I	$[H_2O_2]$	pH	Time	$[Fe^{2+}]/[H_2O_2]$	$0.145 < [Fe^{2+}]/[H_2O_2] < 0.475$ $9 < Time < 12$	$1900 < VSS_{out} < 2100$
I	Time	pH	$[H_2O_2]$	$[Fe^{2+}]/[H_2O_2]$	$0.5 < [Fe^{2+}]/[H_2O_2] < 0.9$ $1220 < [H_2O_2] < 1620$	$1100 < VSS_{out} < 1500$
$[H_2O_2]$	Time	pH	I	$[Fe^{2+}]/[H_2O_2]$	$0.4 < [Fe^{2+}]/[H_2O_2] < 0.8$ $600 < I < 860$	$1400 < VSS_{out} < 1600$
I	$[Fe^{2+}]/[H_2O_2]$	pH	$[H_2O_2]$	Time	$6 < Time < 10$ $770 < [H_2O_2] < 1260$	$500 < VSS_{out} < 1000$
$[H_2O_2]$	$[Fe^{2+}]/[H_2O_2]$	pH	I	Time	$860 < I < 1250$ $7 < Time < 12$	$250 < VSS_{out} < 750$
Time	$[Fe^{2+}]/[H_2O_2]$	pH	I	$[H_2O_2]$	$700 < I < 1000$ $650 < [H_2O_2] < 1320$	$900 < VSS_{out} < 1100$

Artificial Neural Networks Model KSOFM

For the determination of the regularities and correlations between input variables, SOM toolbox (www.cis.hut.fi/projects/somtoolbox) has been used which is applied in MATLAB® (Vesanto et al., 1999). Due to different magnitudes of inputs, all variables were normalized in which they were scaled with the variance of 1 and the mean of 0 (Badalians Gholikandi et al., 2014a). The map size during training was 7×5. Final quantization and topographic error were 0.936 and 0.0, respectively.

The most important features of KSOFM neural network are its visualization which plays an important role in the analyzing results based upon the knowledge extraction (Hong et al., 2003). One of the main kinds of visualization is component plane which shows the value of each neuron to estimate the data variable to the input space using a color bar as code (Grieu et al., 2006; Garcia and Gonzalez 2004). In each component, each hexagon corresponds to the same one in other component planes (Grieu et al., 2006; Garcia and Gonzalez 2004). On the right side of each plane, there is a legend which indicates the range of each variable.

Detecting data clusters is much easier by combining KSOFM model and the Unified Distance Matrix (U-matrix). The idea of the U-matrix is to visualize the feature Map's topology.

In other words, it contains geometrical approximation of the vector distribution in the Kohonen set (Ultsch et al., 1995). Figure 8 illustrates the component planes.

Taguchi Method

In this study, S/N ratio was calculated by equation 5, firstly and the results have shown in the table 10. In the next step, analysis of mean (ANOM) was carried out by using equation 6 and the results are represented in the table 11. Figures 9, 10, 11, and 12 are obtained from the table 11.

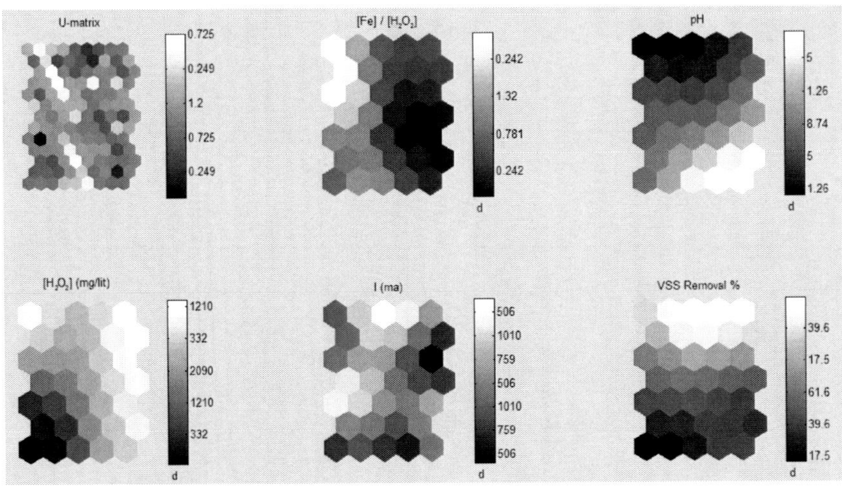

Figure 8. Visualization results of the KSOFM.

Table 10. S/N ratios of the experimental trial results

Trial	VSS removal (%)			S/N Ratio
	Sample 1	Sample 2	Sample 3	
1	9	12	11	20.369
2	37	35	36	31.119
3	25	28	26	28.381
4	15	16.5	17	24.134
5	66	62	63	36.069
6	29	26	17	28.707
7	36.5	36	38	31.317
8	35	37	36	31.119
9	62	61	61.5	35.776
10	66	68	67	36.519
11	27.5	29	30	29.181
12	22.5	24	28	27.792
13	55	52	54	34.586
14	38	35.5	35	31.149
15	40	35	41	31.682
16	13	12	12.5	21.924

Table 11. Mean S/N ratio

Number	Factors	Level 1	Level 2	Level 3	Level 4	L2-L1
1	$[Fe]/[H_2O_2]$	26	31.8	32.32	29.83	5.8
2	pH	31.7	31.87	30.15	26.24	0.17
3	$[H_2O_2]$	25.05	31.67	31.61	31.64	6.62
4	Current density	27.66	31.5	30.72	30.1	3.84

$[Fe]/[H_2O_2]$

Figure 9 represents the maximum removal efficiency of VSS at ratio of $[Fe]/[H_2O_2]=0.58$. By increasing the ratio of $[Fe]/[H_2O_2]$ from level 3 (0.58) to level 4 (1.45) the system efficiency reduces because of reactions 16 and 17 (Umar et al., 2011). Based on reactions 16 and 17 (Umar et al., 2011), high concentrations of iron leads to use hydroxyl radical (OH^0) and hydroperoxyl radical (HO_2^0) as organic oxidizing agents.

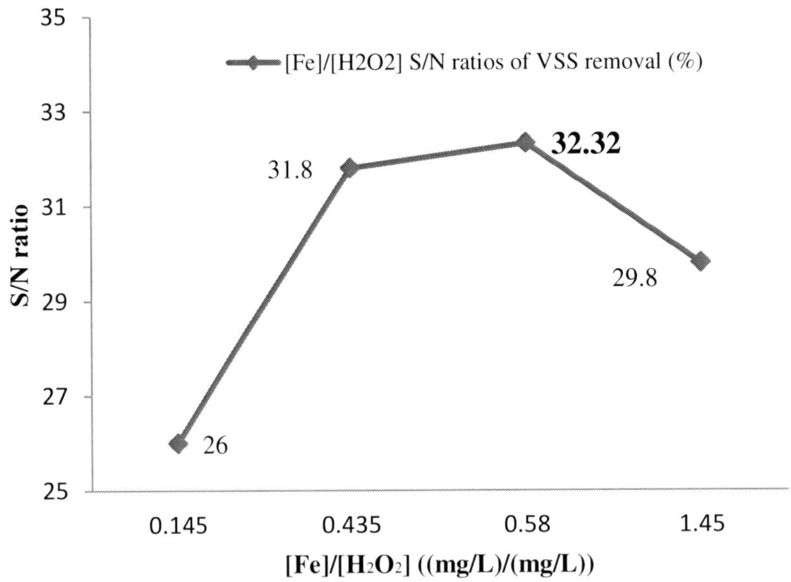

Figure 9. The effect of [Fe]/[H_2O_2] S/N ratios on VSS removal.

$$OH^0 + Fe^{2+} \rightarrow Fe^{3+} + OH^- \quad (16)$$

$$Fe^{3+} + HO_2^0 + H^+ \rightarrow Fe^{2+} + O_2H^+ \quad (17)$$

pH Value

The pH is one of the most important factors in Fenton process. Generally, the Fenton reaction occurs in acidic medium. Previous studies indicated the optimal pH value for the Fenton reaction equals to around 3 (Zhou et al., 2007; Ghoneim et al., 2011).

Figure 10 shows that the highest S/N ratio occurs at level 2, indicating that the maximum removal efficiency of VSS is obtained at pH=3. By increasing the pH value from level 2 to level 3 (pH=5) trace changes will be observed but the pH increments from level 3 to level 4 (pH = 9) causes a sharp decline in the removal efficiency of VSS due to instability of peroxide hydrogen and its conversion into water and oxygen (Wang and Lemley 2001; Shemer and Linden 2006).

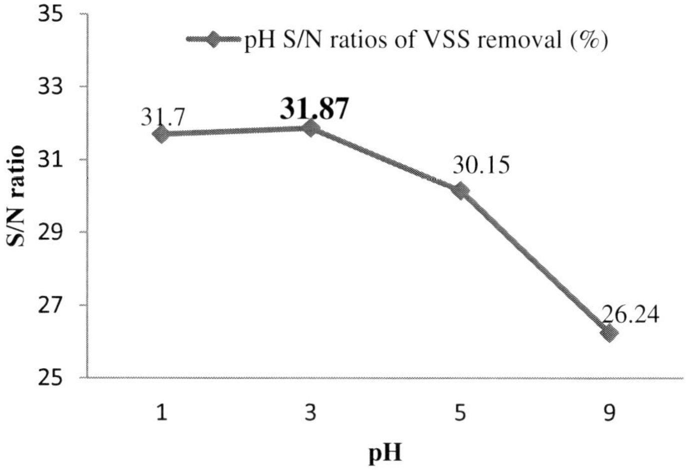

Figure 10. The effect of pH S/N ratios on the VSS removal.

H_2O_2 Concentration

An increase in the concentrations of hydrogen peroxide based on reaction 18 causes the increasing production of hydroxyl radical as the main oxidation agent of organics. Figure 11 shows that maximum value of VSS removal efficiency occurs at level 2. Moreover, hydrogen peroxide concentration variation from level 2 to level 3 (1568mg/L), and then to level 4 (2196mg/L) has insignificant impact on the VSS removal efficiency.

$$Fe^{2+} + H_2O_2 \rightarrow Fe^{3+} + OH^0 + OH^- \qquad 53-76M^{-1}S^{-1} \qquad (18)$$

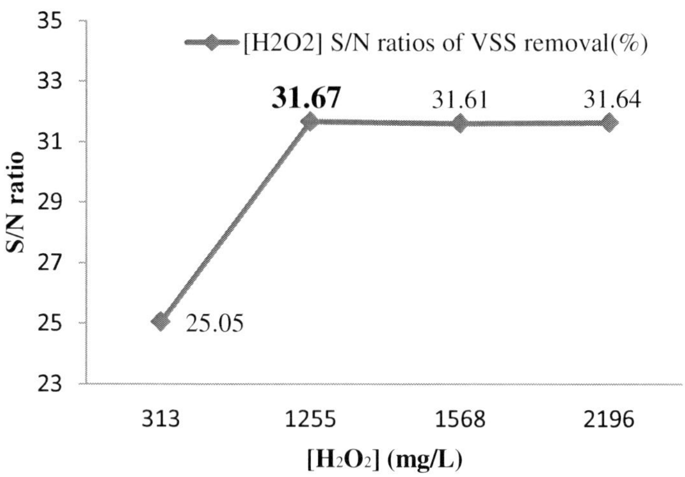

Figure 11. The effect of $[H_2O_2]$ S/N ratios on the VSS removal.

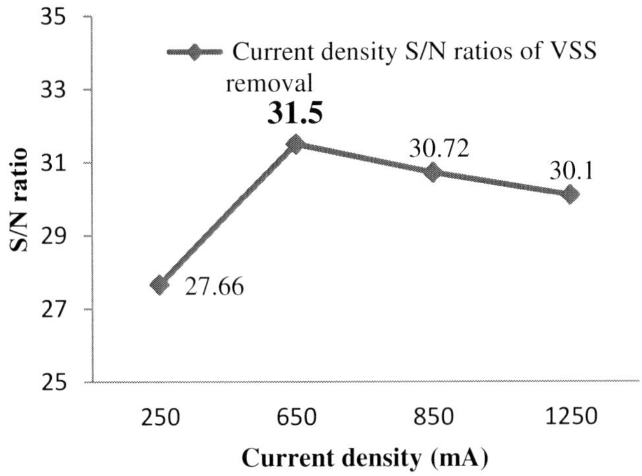

Figure 12. The effect of Current Density S/N ration on the VSS removal.

Current Density

Increasing current density causes an increment in conversion of Fe^{3+} to Fe^{2+} which results in the increasing influence of Fenton process (Wang and Lemley 2001). Figure 12 shows that the VSS removal efficiency has the maximum value at level 2. Due to evolution of hydrogen at the cathode, current densities more than 650mA have inhibiting effect on the reaction 3 (Zhang et al., 2007; Badalians Gholikandi et al., 2014a). Therefore, by increasing this parameter from level 2 to level 3 and then level 4, the VSS removal efficiency will decrease.

ANOVA Analysis

By using ANOVA analysis, error values, "F" variance ratio and contribution percentage of each data will be calculated. The equations 5 to 12 are used accordingly.In this study, Qualitek-4 (w32b) software was employed for the calculation of ANOVA variance, and the results are presented in the table 12. High errors in ANOVA analysis indicate that either the experimental results have high error values or there are other effective parameters affecting the dependent variable, which have not been considered in the design and analysis of experiments. In this study, the amount of error is 0.247%, which is acceptable. The variance ratio (F-Ratio) was calculated for effective parameters and compared with the F values extracted from statistical handbooks tables. If the F-ratio is larger than the F-test, it represents the impact of the parameter on the dependent variable (Roy, 1990). The results of this comparison with determined confidence level are presented in the table 13. With a 99.5% confidence level, it can be said that all considered effective parameters have impact on the dependent variable. The ANOVA analysis also determined the contribution percentage of each parameter. The most effective parameter in this study was the hydrogen peroxide. Also, the optimum conditions likely to provide the maximum VSS removal efficiency are presented in the table 14. As it can be seen, the estimated VSS removal percentage rate (74.05%) has a 2% difference with the tested value (76%).

Table 12. Analysis of Variance (ANOVA)

Number	Factors	Degree Of Freedom	Sums Of Squares	Variance	F-Ratio	Pure sum	Percent (%)
1	[Fe]/ [H_2O_2]	3	98.56	32.853	583.6	98.4	28.573
2	pH	3	82.16	27.385	486.5	81.9	23.809
3	[H2O2]	3	130.37	43.457	771.9	130.2	37.812
4	Current density (mA)	3	33.1	11.028	195.9	32.9	9.559
	other/ Error	3	0.68	0.056			0.247
	Total	15	344.346				100

Table 13. Comparison of the F-ratio and the F-test statistic

Number	Factors	F-Ratio	F-test statistic	Confidence levels
1	[Fe]/[H_2O_2]	583.6	47.46	99.5
2	pH	486.5	47.46	99.5
3	[H_2O_2](mg/L)	771.9	47.46	99.5
4	Current density (mA)	195.9	47.46	99.5

Table 14. The optimum conditions and the performance

Number	Factors	Level Description	Level
1	[Fe]/[H2O2]	0.58	3
2	pH	3	2
3	[H2O2](mg/L)	1255	2
4	Current density (mA)	650	2
Taguchi Result at Optimum Condition			74.05%
Experimental Result at this Condition			76%

Table 15. Optimum condition and performance

Number	Factors	Level Description	Level
1	[Fe]/[H2O2]	0.58	3
2	pH	3	2
3	[H2O2](mg/L)	2196	4
4	Current density (mA)	650	2
ANN Result at Optimum Condition			71.85%
Experimental Result at this Condition			69.50%

A Comparison of the Investigated Modeling Approaches

The Comparison of the Taguchi Method and the Artificial Neural Networks Model MLP

The aim of a well-trained ANN model is to predict expected outputs from introduced inputs (Ongen et al., 2013) which are located within the range of training range not only the data which were in training data (Delnavaz et al., 2010).

The four parameters which were considered as effective ones were described in 4 levels. Therefore, all number of experimental conditions is $4^4 = 256$. For determining the optimum performance (conditions), the ANN-model was fed with conditions which were not examined through 256 data. The highest VSS removal term (condition) was obtained that is shown in the table 15.

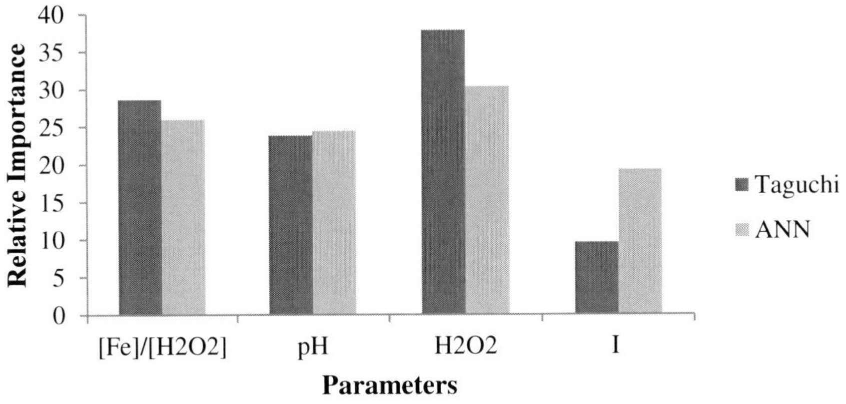

Figure 13. The relative importance comparison of the Taguchi method and the ANN model.

The contribution percentages of the effective parameters were estimated by Taguchi method with Neural Networks. The related results were presented by the figure 13. Both models have identified hydrogen peroxide as the most important effective parameter. The estimated contribution percentage for pH and [Fe]/[H_2O_2] has close results (the difference is below 3%). The maximum difference occurred in the estimation of contribution percentage in current density parameter (9% difference). Comparing the tables 8 and 15 indicates that all parameters were similar in estimated optimal conditions except hydrogen peroxide. In general, both modeling methods are suitable for this system but the Taguchi method is more accurate than neural network method.

References

Aber, S., Salari, D. & Parsa, M. R. (2010). Employing the Taguchi method to obtain the optimum conditions of coagulation–flocculation process in tannery wastewater treatment, *Chemical Engineering Journal*, *162*(1), Elsevier, 127-134.

Abu Qdais, H., Bani Hani K. & Shatnawi N. (2010). Modeling and optimization of biogas production from a waste digester using artificial neural network and genetic algorithm. *Resources, Conservation and Recycling*, *54*(6), 359-363.

Aghav, R. M., Kumar, S. & Mukherjee, S. N. (2011). Artificial neural network modeling in competitive adsorption of phenol and resorcinol from water environment using some carbonaceous adsorbents, *Journal of Hazardous Materials*, *188*(1), Elsevier, 67-77.

Ahmadi, M., Vahabzadeh, F., Bonakdarpour, B., Mofarrah, E. & Mehranian, M. (2005). Application of the central composite design and response surface methodology to the advanced treatment of olive oil processing wastewater using Fenton's peroxidation, *Journal of Hazardous Materials*, *123*(1), Elsevier, 187-195.

Antonopoulou, M., Papadopoulos, V. & Konstantinou, I. (2012). Photocatalytic oxidation of treated municipal wastewaters for the removal of phenolic compounds: optimization and modeling using response surface methodology (RSM) and artificial neural networks (ANNs), *Journal of Chemical Technology and Biotechnology*, *87*(10), Society of Chemical Industry (SCI), Wiley, 1385-1395.

Asghari, A., Kamalabadi, M. & Farzinia, H. (2012). Electrochemical Removal of Methylene Blue from Aqueous Solutions Using Taguchi Experimental Design, *Chemical and Biochemical Engineering Quarterly*, *26*(2), Croatian Society of Chemical Engineers, 145-154.

Atkinson, D. C., Weeks, D. C. & Noll, J. (2004). The design of evolutionary process modeling languages, In: IEEE Software Engineering Conference, 11th Asia-Pacific, 73-82.

Azimipour, M., Badalians Gholikandi, G. & Masihi, H. (2014). Biogas Production with Emphasis on Methanogenesis Process: An Investigative View to Applications of Artificial Neural Networks Modeling, In: Methanogenesis, Biochemistry, Ecological Functions, Natural and Engineered Environments, G. Badalians Gholikandi (Editor), NOVA Science Publishers, New York, 113-148.

Badalians Gholikandi, G., Delnavaz, M. & Riahi, R. (2011). Use of Artificial Neural Network for Prediction of Coagulation/Flocculation Process by PAC in Water Treatment Plant, *Environmental Engineering and Management Journal*, *10*(11), 1719-1725.

Badalians Gholikandi, G. & Khosravi, M. (2012). Upgrading of Submerged Membrane Bioreactor Operation with regard to Soluble Microbial Products and Mathematical Modeling for Optimization of Critical Flux, *Desalination and Water Treatment*, *39*(1-3), Taylor & Francis, 199-208.

Badalians Gholikandi, G., Masihi, H., Azimipour, M., Abrishami, A. & Mirabi, M. (2014a). Optimizing stabilization of waste-activated sludge using Fered-Fenton process and artificial neural network modeling (KSOFM, MLP), *Journal of Environmental Science and Pollution Research*, *21*(11), Springer, 7177-7186.

Badalians Gholikandi, G., Jamshidi, S. & Hazrati, H. (2014b). Optimization Of Anaerobic Baffled Reactor (ABR) Using Artificial Neural Network In Municipal Wastewater Treatment, *Environmental Engineering and Management Journal*, *13*(1), 95-104.

Banks, J., Carson, S. J., Nelson, B. N. & Nicol, D. M. (2005). Discrete-event system simulation, 4th ed., New Jersey, Prentice-Hall, Inc.

Benítez, J., Rodriguez, A. B. R. A. H. A. M. & Suárez, A. (1994). Optimization technique for sewage sludge conditioning with polymer and skeleton builders, Water Research 28(10), *the International Water Association (IWA)*, Elsevier, 2067-2073.

Bhatti, M. S., Kapoor, D., Kalia, R. K., Reddy, A. S. & Thukral, A. K. (2011). RSM and ANN modeling for electrocoagulation of copper from simulated wastewater: Multi objective optimization using genetic algorithm approach, *Desalination*, *274*(1), Elsevier, 74-80.

Bolanča, T., Ukić, Š., Peternel, I., Kušić, H. & Božić, A. L. (2014). Artificial neural network models for advanced oxidation of organics in water matrix-Comparison of applied methodologies, *Indian Journal of Chemical Technology*, *21*(1), 21-29.

Boniecki, P., Dach, J., Pilarski, K. & Piekarska-Boniecka, H. (2012). Artificial neural networks for modeling ammonia emissions released from sewage sludge composting, *Atmospheric Environment*, *57*, Elsevier, 49-54.

Capodaglio, A. G., Jones, H. V., Novotny, V. & Feng, X. (1991). Sludge bulking analysis and forecasting: application of system identification and artificial neural computing technologies, *Water Research*, *25*(10), the International Water Association (IWA), Elsevier, 1217-1224.

Chan, C. W. & Huang, G. H. (2003). Artificial intelligence for management and control of pollution minimization and mitigation processes, *Engineering Applications of Artificial Intelligence*, *16*(2), the International Federation of Automatic Control (IFAC), Elsevier, 75-90.

Chen, J. C., Chang, N. B. & Shieh, W. K. (2003). Assessing wastewater reclamation potential by neural network model, Engineering applications of artificial intelligence, 16(2), 149-157.

Chen, S. H., Jakeman, A. J. & Norton, J. P. (2008). Artificial intelligence techniques: an introduction to their use for modelling environmental systems, *Mathematics and Computers in Simulation*, *78*(2), International Association for Mathematics and Computers in Simulation, Elsevier, 379-400.

Choi, D. J. & Park, H. (2001). A hybrid artificial neural network as a software sensor for optimal control of a wastewater treatment process, *Water Research*, *35*(16), the International Water Association (IWA), Elsevier, 3959-3967.

Çinar, Ö. (2005). New tool for evaluation of performance of wastewater treatment plant: artificial neural network, *Process Biochemistry*, *40*(9), Elsevier, 2980-2984.

Cortes, U., Sanchez-Marre, M., Ceccaroni, L., R-Roda, I. & Poch, M. (2000). Artificial intelligence and environmental decision support systems, *Applied Intelligence*, *13*(1), Springer, 77-91.

Dayhoff, J. E. (1990). Neural network architectures: an introduction, Van Nostrand Reinhold Co.

de Freitas, A. P. B. R., de Freitas, L. V., Loures, C. C. A., dos Santos, M. A. R., Ricardo, G. D., Marins, F. A. S., dos Santos, H.T.L., Samanamud, G.L., Amaral, M.S. & Silva, M. B. (2013). Taguchi Method Applied to Environmental Engineering, In: Design of Experiments-Applications, INTEC, 80-92.

de Freitas, L. V., de Freitas, A. P. B. R., Marins, F. A. S., Loures, C. C. A., Silva, M. B., Salman, F., dos Santos, H. T. L. & Samanamud, G. L. (2013). Multivariate Analysis in Advanced Oxidation Process, INTEC.

Dellana, S. A. & West, D. (2009). Predictive modeling for wastewater applications: Linear and nonlinear approaches, *Environmental Modelling & Software*, *24*(1), Elsevier, 96-106.

Delnavaz, M., Ayati, B. & Ganjidoust, H. (2010). Prediction of moving bed biofilm reactor (MBBR) performance for the treatment of aniline using artificial neural networks (ANN), *Journal of Hazardous Materials*, *179*, Elsevier, 769-775.

Domonkos, T. (2010). Computer Simulation as a Tool for Analyzing and Optimizing Real-Life Processes, *Management Information Systems*, *5*(1), 13-18.

Elmolla, E. S., Chaudhuri, M. & Eltoukhy, M. M. (2010). The use of artificial neural network (ANN) for modeling of COD removal from antibiotic aqueous solution by the Fenton process, *Journal of Hazardous Materials*, *179*, Elsevier, 127-134.

Engin, A. B., Özdemir, Ö., Turan, M. & Turan, A. Z. (2008). Color removal from textile dye bath effluents in a zeolite fixed bed reactor: Determination of optimum process conditions using Taguchi method, *Journal of Hazardous Materials*, *159*(2), Elsevier, 348-353.

Fang, F., Ni, B. J., Xie, W. M., Sheng, G. P., Liu, S. G., Tong, Z. H. & Yu, H. Q. (2010). An integrated dynamic model for simulating a full-scale municipal wastewater treatment plant under fluctuating conditions, *Chemical Engineering Journal*, *160*(2), Elsevier, 522-529.

Farzadkia, M., Dehghani, M. & Moafian, M. (2014). The effects of Fenton process on the removal of petroleum hydrocarbons from oily sludge in Shiraz oil refinery, Iran, *Journal of Environmental Health, Science and Engineering*, *12*(1), 31.

Gall, B. (1999). Review of activated sludge modelling, In: TAPPI International Environmental Conference and Exhibit, Nashville, TN, USA.

Garcia, H. & Gonzalez, L. (2004). Self-Organizing Map and Clustering for Wastewater Treatment Monitoring, *Engineering Applications of Artificial Intelligence*, *17*(3), the International Federation of Automatic Control (IFAC), Elsevier, 215-225.

Ghafari, S., Aziz, H. A., Isa, M. H. & Zinatizadeh, A. A. (2009). Application of response surface methodology (RSM) to optimize coagulation–flocculation treatment of leachate using poly-aluminum chloride (PAC) and alum, *Journal of Hazardous Materials*, *163*(2), Elsevier, 650-656.

Ghaseminezhad, M. H., Karami, A. (2011). A novel self-organizing map (SOM) neural network for discrete groups of data clustering, Applied Soft Computing 11(4), Elsevier, 3771-3778.

Ghoneim, M. M., El-Desoky, H. S. & Zidan, N. M. (2011). Electro-Fenton oxidation of Sunset Yellow FCF azo-dye in aqueous solutions, *Desalination*, *274*, Elsevier, 22–30.

Göb, S., Oliveros, E., Bossmann, S. H., Braun, A. M., Guardani, R. & Nascimento, C. A. (1999). Modeling the kinetics of a photochemical water treatment process by means of artificial neural networks, *Chemical Engineering and Processing: Process Intensification*, *38*(4), Elsevier, 373-382.

Gontarski, C. A., Rodrigues, P. R., Mori, M. & Prenem, L. F. (2000). Simulation of an industrial wastewater treatment plant using artificial neural networks, *Computers & Chemical Engineering*, *24*(2), Elsevier, 1719-1723.

Grieu S., Thiery F., Traore A., Nguyen T. P., Barreau M. & Polit M. (2006). KSOM and MLP neural networks for on-line estimating the efficiency of an activated sludge process, *Chemical Engineering Journal*, *116*, 1–11.

Güclü, D., Yilmaz, N. & Ozkan-Yucel, U. G. (2011). Application of neural networks prediction model to full-scale anaerobic sludge digestion, *Journal of Chemical Technology and Biotechnology*, 86, Society of Chemical Industry (SCI), Wiley, 691-698.

Gujer, W., Henze, M., Mino, T. & Loosdrecht, M. V. (1999). Activated sludge model no. 3, *Water Science and Technology*, *39*(1), IWA Publishing, 183-193.

Gujer, W., Henze, M., Mino, T., Matsuo, T., Wentzel, M. C. & Marais, G. V. R. (1995). The activated sludge model No. 2: biological phosphorus removal, *Water Science and Technology*, *31*(2), IWA Publishung, 1-11.

Hamed, M. M., Khalafallah, M. G. & Hassanien, E. A. (2004). Prediction of wastewater treatment plant performance using artificial neural networks, *Environmental Modelling & Software*, *19*(10), Elsevier, 919-928.

Hasanlou, H., Mehrdadi, N., Jafarzadeh, M. T., & Hasanlou, H. (2012). Performance Simulation of H-TDS Unit of Fajr Industrial Wastewater Treatment Plant Using a Combination of Neural Network and Principal Component Analysis, *Journal of Water Resource & Protection*, *4*(5).

Haykin, S. & Network, N. (2004). A comprehensive foundation, *Neural Networks*, *2*.

Heikkinen, M., Poutiainen, H., Liukkonen, M., Heikkinen, T. & Hiltunen, Y. (2011). Subtraction analysis based on self-organizing maps for an industrial wastewater treatment process, *Mathematics and Computers in Simulation*, *82*(3), International Association for Mathematics and Computers in Simulation, 450-459.

Henze, M., Grady, C. P. L., Gujer, W., Marais, G. & Matsuo, T. (1987). Activated sludge model No. 1. IAWPRC Task group on mathematical modelling for design and operation of biological wastewater treatment. Scientific and Technical Report.

Holubar, P., Zani, L., Hager, M., Froeschl, W., Radak, Z. & Braun, R. (2002). Advanced controlling of anaerobic digestion by means of hierarchical neural networks, *Water Research*, *36*, the International Water Association (IWA), Elsevier, 2582-2588.

Hong, Y. S. T., Rosen, M. R. & Bhamidimarri, R. (2003). Analysis of a municipal wastewater treatment plant using a neural network-based pattern analysis, *Water Research*, *37*(7), the International Water Association (IWA), Elsevier, 1608-1618.

Jakeman, A. J., Voinov, A. A., Rizzoli, A. E. & Chen, S. H. (Eds.) (2008). Environmental Modelling, Software and Decision Support: State of the art and new perspective, Vol. *3*, Elsevier.

Jahandideh, S., Barzegari Asadabadi, E., Askarian, M., Movahedi, M. M., Hosseini, S. & Jahandideh, M. (2009). The use of artificial neural networks and multiple linear regression to predict rate of medical waste generation, *Waste Management*, *29*, Elsevier, 2874-2879.

Kasiri, M. B., Aleboyeh, H. & Aleboyeh, A. (2008). Modeling and optimization of heterogeneous photo-Fenton process with response surface methodology and artificial neural networks, *Environmental Science & Technology*, *42*(21), American Chemical Society (ACS), 7970-7975.

Kanat, G. & Saral, A. (2009). Estimation of biogas production rate in a thermophilic UASB reactor using artificial neural networks, *Environmental Modeling & Assessment*, *14*(5), Springer, 607-614.

Kavitha, S., Adish Kumar, S., Yogalakshmi, K. N., Kaliappan, S. & Rajesh Banu, J. (2013). Effect of enzyme secreting bacterial pretreatment on enhancement of aerobic digestion potential of waste activated sludge interceded through EDTA, *Bioresource Technology*, *150*, Elsevier, 210-219.

Khandan, N. (2010). Modeling tools for environmental engineers and scientists, CRC Press.

Khataee, A. R. & Kasiri, M. B. (2010). Artificial neural networks modeling of contaminated water treatment processes by homogeneous and heterogeneous nanocatalysis, *Journal of Molecular Catalysis A: Chemical*, *331*(1), Elsevier, 86-100.

Körbahti, B. K., Aktaş, N. & Tanyolaç, A. (2007). Optimization of electrochemical treatment of industrial paint wastewater with response surface methodology, *Journal of Hazardous Materials*, *148*(1), Elsevier, 83-90.

Körbahti, B. K. (2007). Response surface optimization of electrochemical treatment of textile dye wastewater, *Journal of Hazardous Materials*, *145*(1), Elsevier, 277-286.

Körbahti, B. K. & Tanyolaç, A. (2008). Electrochemical treatment of simulated textile wastewater with industrial components and Levafix Blue CA reactive dye: optimization through response surface methodology, *Journal of Hazardous Materials*, *151*(2), Elsevier, 422-431.

Lee, D. S., Jeon, C. O., Park, J. M. & Chang, K. S. (2002). Hybrid neural network modeling of a full scale industrial wastewater treatment process, *Biotechnology and bioengineering*, *78*(6),Wiley, 670-682.

Lee, B. H. & Scholz, M. (2006). Application of the self-organizing map (SOM) to assess the heavy metal removal performance in experimental constructed wetlands, *Water Research*, *40*(18), the International Water Association (IWA), Elsevier, 3367-3374.

Lee, J. W., Suh, C., Hong, Y.S.T., & Shin, H.S. (2011). Sequential modeling of a full-scale wastewater treatment plant using an artificial neural network, *Bioprocess and Biosystems Engineering*, *34*(8), Springer, 963-973.

Lin, C. H., Yu, R. F., Cheng, W. P. & Liu, C. R. (2012). Monitoring and control of UV and UV-TiO< sub> 2</sub> disinfections for municipal wastewater reclamation using artificial neural networks, *Journal of Hazardous Materials*, *209*, Elsevier, 348-354.

Llorens, E., Thiery, F., Grieu, S. & Polit, M. (2008). Evaluation of WWTP discharges into a Mediterranean river using KSOM neural networks and mass balance modelling, *Chemical Engineering Journal*, *142*(2), Elsevier, 135-146.

Lou, I. & Zhao, Y. (2012). Sludge bulking prediction using principle component regression and artificial neural network, Mathematical Problems in Engineering, Hindawi Publishing Co.

Loures, C. C., Filho, H. J. I., Samanamud, G. R. L. & Souza, A. L. (2013). Performance Evaluation of Photo-Fenton and Fenton Processses for Dairy Effluent Treatment, *International Review of Chemical Engineering-Rapid Communications (IRECHE)*, *5*(4), Praise Worthy Prize.

Madaeni, S. S. & Koocheki, S. (2006). Application of taguchi method in the optimization of wastewater treatment using spiral-wound reverse osmosis element, *Chemical Engineering Journal*, *119*(1), Elsevier, 37-44.

McCulloch, W. S. & Pitts, W. (1990). A logical calculus of the ideas immanent in nervous activity, *Bulletin of Mathematical Biology*, *52*(1), Springer, 99-115.

Mehrdadi, N., Hasanlou, H., Jafarzadeh, M. T. & Abdolabadi, H. (2012). Simulation of Low TDS and Biological Units of Fajr Industrial Wastewater Treatment Plant Using Artificial Neural Network and Principal Component Analysis Hybrid Method, *Journal of Water Resource & Protection*, *4*(6), Science Research Publ..

Mjalli, F. S., Al-Asheh, S. & Alfadala, H. E. (2007). Use of artificial neural network black-box modeling for the prediction of wastewater treatment plants performance, *Journal of Environmental Management*, *83*(3), Elsevier, 329-338.

Mohammadi, T. & Safavi, M. A. (2009). Application of Taguchi method in optimization of desalination by vacuum membrane distillation, *Desalination*, *249*(1), Elsevier, 83-89.

Moral, H., Aksoy, A. & Gokcay, C. F. (2008). Modeling of the activated sludge process by using artificial neural networks with automated architecture screening, *Computers & Chemical Engineering*, *32*(10), Elsevier, 2471-2478.

Mourabet, M., El Rhilassi, A., Bennani-Ziatni, M. & Taitai, A. (2014). Comparative Study of Artificial Neural Network and Response Surface Methodology for Modelling and Optimization the Adsorption Capacity of Fluoride onto Apatitic Tricalcium Phosphate, *Universal Journal of Applied Mathematics*, *2*(2), Horizon Research Publishing Co., 84-91.

Mustafa, Y. A., Jaid, G. M., Alwared, A. I. & Ebrahim, M. (2014). The use of artificial neural network (ANN) for the prediction and simulation of oil degradation in wastewater by AOP, *Journal of Environmental Science and Pollution Research*, Springer, 1-8.

Nasr, M. S., Moustafa, M. A., Seif, H. A. & El Kobrosy, G. (2012). Application of Artificial Neural Network (ANN) for the prediction of EL-AGAMY wastewater treatment plant performance-EGYPT, *Alexandria Engineering Journal*, *51*(1), 37-43.

Qiao, J. & Wang, H. (2008). A self-organizing fuzzy neural network and its applications to function approximation and forecast modeling, *Neurocomputing*, *71*(4), Elsevier, 564-569.

Ongen, A., Ozcan, H. K. & Arayıcı, S. (2013). An evaluation of tannery industry wastewater treatment sludge gasification by artificial neural network modeling, *Journal of Hazardous Materials*, *263*, Elsevier, 361-366.

Ozkaya, B., Demir, A. & Bilgili, M. S. (2007). Neural network prediction model for the methane fraction in biogas from field-scale landfill bioreactors, *Environmental Modelling & Software*, *22*(6), Elsevier, 815-822.

Patry, G. G. & Chapman, D. T. (1989). Dynamic modeling and expert systems in wastewater engineering, CRC Press, Inc..

Pendashteh, A. R., Fakhru'l Razi, A., Chaibakhsh, N., Chuah Abdullah, L., Madaeni, S. S. & Zainal Abidin, Z. (2011). Modeling of membrane bioreactor treating hypersaline oily wastewater by artificial neural network, *Journal of Hazardous Materials*, *192*, Elsevier, 568-575.

Rahmanian, B., Pakizeh, M., Mansoori, S. A. A. & Abedini, R. (2011). Application of experimental design approach and artificial neural network (ANN) for the determination of potential micellar-enhanced ultrafiltration process, *Journal of Hazardous Materials*, *187*, Elsevier, 67-74.

Radermacher, F. J., Riekert, W. F., Page, B. & Hilty, L. M. (1994). Trends in environmental information processing, In: *Proceedings of the IFIP Congress*, Vol. *2*, R. Raubold, K. Brunnstein (eds.), Elsevier.

Raduly, B., Gernaey, K. V., Capodaglio, A. G., Mikkelsen, P. S. & Henze, M. (2007). Artificial neural networks for rapid WWTP performance evaluation: Methodology and case study, *Environmental Modelling & Software*, *22*(8), Elsevier, 1208-1216.

Ramsey, A. L. (1994). Assessment of the modeling abilities of neural networks, Doctoral dissertation, Massachusetts Institute of Technology.

Rao, R. S., Kumar, C. G., Prakasham, R. S. & Hobbs, P. J. (2008). The Taguchi methodology as a statistical tool for biotechnological applications: a critical appraisal, *Biotechnology journal*, *3*(4), Wiley, 510-523.

Rastegar, S. O., Mousavi, S. M., Shojaosadati, S. A. & Sheibani, S. (2011). Optimization of petroleum refinery effluent treatment in a UASB reactor using response surface methodology, *Journal of Hazardous Materials*, *197*, Elsevier, 26-32.

Rosenblatt, F. (1958). The perceptron: a probabilistic model for information storage and organization in the brain, *Psychological Review*, *65*(6), American Psychological Association, 386.

Roy, R. (1990). A Primer on the Taguchi Method, First Ed., Society of Manufacturing Engineers, New York.

Rumelhart, D. E., Hinton, G. E. & Williams, R.J . (1986). Learning representations by back-propagating errors, *Nature*, *323*, 533 – 536.

Rustum, R., Adeloye, A. J. & Scholz, M. (2008). Applying Kohonen Self-Organizing Map as a Software Sensor to Predict Biochemical Oxygen Demand, *Water Environment Research*, *80*, Water Environment Federation (WEF), 32-40.

Saien, J., Soleymani, A. R. & Bayat, H. (2012). Modeling fentonic advanced oxidation process decolorization of Direct Red 16 using artificial neural network technique, *Desalination and Water Treatment*, *40*(1-3), Taylor & Francis, 174-182.

Salari, D., Daneshvar, N., Aghazadeh, F. & Khataee, A. R. (2005). Application of artificial neural networks for modeling of the treatment of wastewater contaminated with methyl tert-butyl ether (MTBE) by UV/H_2O_2 process, *Journal of Hazardous Materials*, *125*, Elsevier, 205–210.

Schichl, H. (2004). Models and the history of modeling, In: Modeling Languages in Mathematical Optimization, Springer, 25-36.

Shemer, H. & Linden, K. G. (2006). Degradation and by-product formation of diazinon in water during UV and UV/H_2O_2 treatment, *Journal of Hazardous Materials*, *136*, Elsevier, 553-559.

Silva, M. B., Guimarães, O. L. C., Francisco, A., Siqueira Filho, H. J. I., Villela Filho, D. N., de Aquino, H. O. Q., ... & de Oliveira Almeida, C. R. (2011). Applications of Neural Networks in Advanced Oxidative Process, Artificial Neural Networks - Application, Dr. Chi Leung Patrick Hui(Ed.),http://www.intechopen.com/books/artificial-neural-networks-application/applications-of-neural-networks-in-advanced-oxidative-process.

Soloman, P. A., Ahmed Basha, C., Velan, M., Balasubramanian, N. & Marimuthu, P. (2009). Augmentation of biodegradability of pulp and paper industry wastewater by electrochemical pre-treatment and optimization by RSM, *Separation and Purification Technology*, *69*(1), Elsevier, 109-117.

Syu, M., Chen, B. & Chou, S. (2003). A study on the sedimentation model and neural network online adaptive control of a benzoic acid imitated wastewater oxidation process, *Industrial & Engineering Chemistry Research*, *42*(26), American Chemical Society (ACS), 6862-6871.

Taguchi, G. (1990). Introduction to Quality Engineering, McGraw-Hill, New York, USA, 191.

Taheri, M., Moghaddam, M. R. A., & Arami, M. (2012). Optimization of Acid Black 172 decolorization by electrocoagulation using response surface methodology, *Iranian journal of environmental health science & engineering*, *9*(1), 1-8.

Tashaouie, H. R., Badalians Gholikandi, G. & Hazrati, H. (2012). Artificial neural network modeling for predict performance of pressure filters in a water treatment plant, *Desalination and Water Treatment*, *39*, Taylor & Francis, 192-198.

Tezel, G., Yel, E. & Sinan, R. K. (2010). Artificial Neural Network (ANN) model for domestic wastewater treatment plant control, In: The fourth scientific conference on water observation and information system for decision support (BALWOIS), Ohrid, Republic of Macedonia.

Tobiszewski, M., Tsakovski, S., Simeonov, V. & Namieśnik, J. (2012). Chlorinated solvents in a petrochemical wastewater treatment plant: An assessment of their removal using self-organising maps, *Chemosphere*, *87*(8), Elsevier, 962-968.

Umar, M., Aziz, H.A. & Suffian Yusoff, M. (2010). Trends in the use of Fenton, electro-Fenton and photo-Fenton for the treatment of landfill leachate, *Waste Management*, *30*, Elsevier, 2113-2121.

Venkata Mohan, S., Chandrasekhara Rao, N., Krishna Prasad, K., Murali Krishna, P., Sreenivas Rao, R. & Sarma, P. N. (2005). Anaerobic treatment of complex chemical wastewater in a sequencing batch biofilm reactor: Process optimization and evaluation of factor interactions using the Taguchi dynamic DOE methodology, *Biotechnology and bioengineering*, *90*(6), Wiley, 732-745.

Vesanto, J., Himberg, J., Alhoniemi, E. & Parhankangas, J. (1999). Self-organizing map in Matlab: the SOM Toolbox, *In Proceedings of the Matlab DSP conference*, Vol. *99*, 16-17.

Vyas, M., Modhera, B. & Sharma, A. K. (2011). Artificial Neural Network Based Model in Effluent Treatment Process, *International Journal of Advanced Engineering Technology*, *2*(2), Blue Eyes Intelligence Engineering & Sciences Publication, 271-275.

Walker, R. (1993). Artificial intelligence, *Paper Maker*, *56* (3), 24–26.

Wang, Q. & Lemley, A. T. (2001). Kinetic model and optimization of 2,4-D degradation by anodic Fenton treatment, *Environmental Science & Technology*, *35*, American Chemical Society (ACS), 4509–4514.

Wang, J. P., Chen, Y. Z., Ge, X. W. & Yu, H. Q. (2007). Optimization of coagulation–flocculation process for a paper-recycling wastewater treatment using response surface methodology, *Colloids and Surfaces A: Physicochemical and Engineering Aspects*, *302*(1), Elsevier, 204-210.

Whiteman, J. K. & Kana, E. G. (2014). Comparative Assessment of the Artificial Neural Network and Response Surface Modelling Efficiencies for Biohydrogen Production on Sugar Cane Molasses, *BioEnergy Research*, *7*(1), Springer, 295-305.

Widrow, B. (1962). Generalization and information storage in network of adaline 'neurons', Self-organizing systems, 435-462.

Wu, Y., Zhou, S., Qin, F., Ye, X. & Zheng, K. (2010). Modeling physical and oxidative removal properties of Fenton process for treatment of landfill leachate using response surface methodology (RSM), *Journal of Hazardous Materials*, *180*(1), Elsevier, 456-465.

Yu, R. F., Chen, H. W., Liu, K. Y., Cheng, W. P. & Hsieh, P. H. (2010). Control of the Fenton process for textile wastewater treatment using artificial neural networks, *Journal of Chemical Technology and Biotechnology*, *85*(2), Society of Chemical Industry (SCI), Wiley, 267-278.

Yu, J., Zhou, S. & Wang, W. (2010). Combined treatment of domestic wastewater with landfill leachate by using A< sup> 2</sup>/O process, *Journal of Hazardous Materials*, *178*(1), Elsevier, 81-88.

Yusof, T. R. T., Man, H. C. & Hafid, H. S. (2014). Optimization of Methane Gas Production From Co-digestion of Food waste and Poultry Manure Using Artificial Neural Network and Response Surface Methodology, *Journal of Agricultural Science*, *6*(7), Canadian Center of Science and Education, 27.

Zeng, G. M., Qin, X. S., He, L., Huang, G. H., Liu, H. L. & Lin, Y. P. (2003). A neural network predictive control system for paper mill wastewater treatment, *Engineering Applications of Artificial Intelligence*, *16*, the International Federation of Automatic Control (IFAC), Elsevier, 121-129.

Zhang, H., Fei, C., Zhang, D. & Tang, F. (2007). Degradation of 4-nitrophenol in aqueous medium by electro-Fenton method, *Journal of Hazardous Materials*, *145*, Elsevier, 227-232.

Zhang, R. & Hu, X. (2012). Effluent Quality Prediction of Wastewater Treatment System Based on Small-world ANN, *Journal of Computers*, *7*(9), Academy Publisher.

Zheng, C., Wang, M., Wang, Y. & Huang, Z. (2012). Optimization of biosurfactant-mediated oil extraction from oil sludge, *Bioresource Technology*, *110*, Elsevier, 338-342.

Zhou, M., Yu, Q., Lei, L. & Barton, G. (2007). Electro-Fenton method for the removal of methyl red in an efficient electrochemical system, *Separation and Purification Technology*, *573*, Elsevier, 380-387.

Zhu, X., Tian, J., Liu, R. & Chen, L. (2011). Optimization of Fenton and electro-Fenton oxidation of biologically treated coking wastewater using response surface methodology, *Separation and Purification Technology*, *81*(3), Elsevier, 444-450.

Zinatizadeh, A. A. L., Mohamed, A. R., Abdullah, A. Z., Mashitah, M. D., Hasnain Isa, M. & Najafpour, G. D. (2006). Process modeling and analysis of palm oil mill effluent treatment in an up-flow anaerobic sludge fixed film bioreactor using response surface methodology (RSM), *Water Research*, *40*(17), the International Water Association (IWA), Elsevier, 3193-3208.

Zirehpour, A., Rahimpour, A., Jahanshahi, M. & Peyravi, M. (2014). Mixed matrix membrane application for olive oil wastewater treatment: Process optimization based on Taguchi design method, *Journal of Environmental Management*, *132*, Elsevier, 113-120.

Zolfaghari, G., Esmaili Sari, A., Anbia, M., Younesi, H., Amirmahmood, S. & Ghafari Nazari, A. (2011). Taguchi optimization approach for Pb(II) and Hg(II) removal from aqueous solutions using modified mesoporous carbon, *Journal of Hazardous Materials*, *192*, Elsevier, 1046-1055.

INDEX

A

B

C

D

E

F

J

K

L

M

N

Q

R

S

T

U

V

W

X

Y

Z